STUDY AND REVISION GUIDE

Cambridge IGCSE™ and O Level

Geography

Revised edition

Paul Guinness
Garrett Nagle

HODDER
EDUCATION
AN HACHETTE UK COMPANY

Dedication:

To Angela (Garrett Nagle)
To Mary (Paul Guinness)

The Publishers would like to thank the following for permission to reproduce copyright material.

Photo credits

p.52 Fig. 2.10 © Mario Hagen - stock.adobe.com; **p.55** Fig. 2.17 © Garrett Nagle; **p.65** Fig. 3.1 © Rubens Alarcon/Getty Images/Hemera/Thinkstock; **p.122** Fig. 4.13 and Fig. 4.15 © Garrett Nagle.

Every effort has been made to trace all copyright holders, but if any have been inadvertently overlooked, the Publishers will be pleased to make the necessary arrangements at the first opportunity.

This text has not been through the Cambridge endorsement process. All exam-style questions and sample answers in this title were written by the authors. In examinations, the way marks are awarded may be different.

Orders: please contact Bookpoint Ltd, 130 Park Drive, Milton Park, Abingdon, Oxon OX14 4SE.
Telephone: +44 (0)1235 827827. Fax: +44 (0)1235 400401. Email education@bookpoint.co.uk. Lines are open from 9 a.m. to 5 p.m., Monday to Saturday, with a 24-hour message answering service. You can also order through our website: www.hoddereducation.com

ISBN: 9781510421394

© Garrett Nagle and Paul Guinness 2019

First published in 2016
This edition published in 2019 by
Hodder Education,
An Hachette UK Company
Carmelite House
50 Victoria Embankment
London EC4Y 0DZ

www.hoddereducation.com

Impression number   10 9 8 7 6 5 4 3 2 1

Year   2020   2019

Cover photo © Shutterstock/Joachim Wendler

Illustrations by Integra Software Services and Barking Dog Art

Typeset in ITC Officina Sans Std 11/13 by Integra Software Services Pvt. Ltd., Pondicherry, India

Printed in India

A catalogue record for this title is available from the British Library.

MIX
Paper from
responsible sources
FSC™ C104740

Contents

Introduction

Welcome to the *Cambridge IGCSE™ and O Level Geography Study and Revision Guide*. This book has been written to help you revise everything you need to know for your Geography exam alongside the *Cambridge IGCSE and O Level Geography Third Edition Student's Book*. Following the Geography syllabus, it covers all the key content as well as sample questions and answers, case studies and practice questions to help you learn how to answer questions and to check your understanding.

How to use this book

Key objectives

A summary of the main information.

Test yourself

Questions for you to check your understanding and progress.

Cross-references to the Student's Book are shown by this icon .

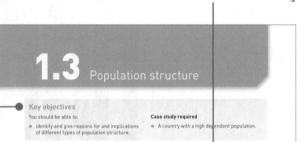

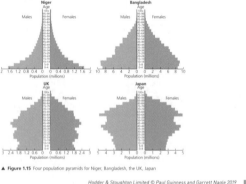

Key definitions

Definitions of the key terms you need to know.

Tip

Advice to help you give the perfect answer.

Cambridge IGCSE and O Level Geography Study and Revision Guide

Sample exam questions

Exam-style questions for you to think about. ————

Student's answers ————

Model student answers to see how the question might be answered.

Teacher's comments

Feedback from an examiner showing what was good, and what could be improved.

Exam-style questions

Exam questions for you to try to see what you have learned.

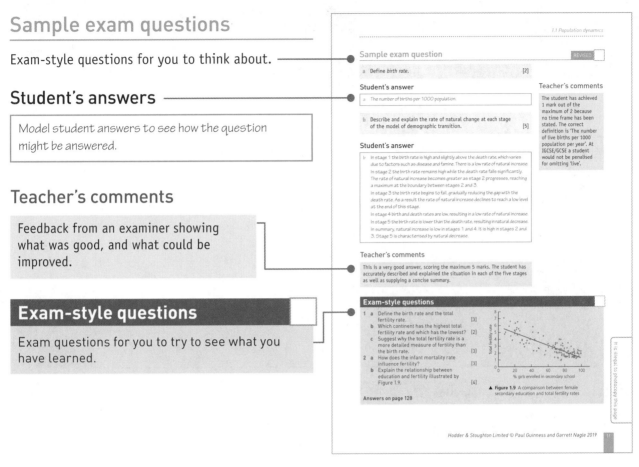

Common error

Mistakes that students often make, and how to avoid them.

Case study

Real examples to help you explain what you have learned.

Answers

Outline answers to the Test yourself questions and the Exam-style questions from page 125.

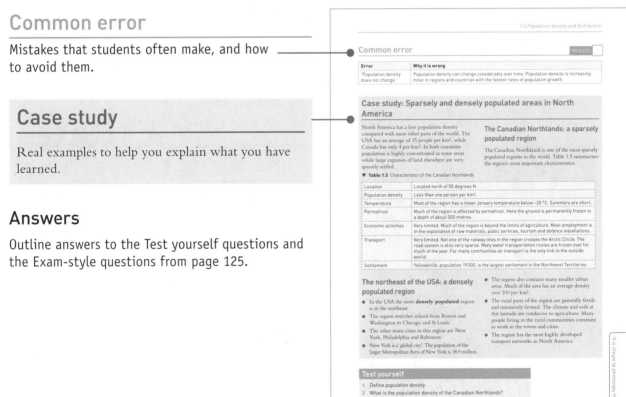

Key objectives

You should be able to:

- describe and give reasons for the rapid increase in the world's population
- show an understanding of over-population and under-population
- understand the main causes of a change in population size
- give reasons for contrasting rates of natural population change
- describe and evaluate population policies.

Case studies required

- A country which is over-populated.
- A country which is under-populated.
- A country with a high rate of natural population growth.
- A country with a low rate of population growth (or population decline).

Key definitions

REVISED

Term	Definition
Population explosion	The rapid population growth of the developing world in the post-1950 period.
Birth rate	The number of live births per 1000 population in a year.
Death rate	The number of deaths per 1000 population in a year.
Rate of natural change	The difference between the birth rate and the death rate. If it is positive it is termed natural increase. If it is negative it is known as natural decrease.
Rate of net migration	The difference between the rates of immigration and emigration.
Model of demographic transition	A model illustrating the historical shift of birth and death rates from high to low levels in a population.
Total fertility rate	The average number of children a women has during her lifetime.
Infant mortality rate	The number of deaths of children under one year of age per 1000 live births per year.
Life expectancy at birth	The average number of years a newborn infant can expect to live under current mortality levels.
Depopulation	A decline in the number of people in a population.
Optimum population	The best balance between a population and the resources available to it. This is usually viewed as the population giving the highest average living standards in a country.
Under-population	When there are too few people in an area to use the resources available effectively.
Over-population	When there are too many people in an area relative to the resources and the level of technology available.
Underemployment	A situation where people are working less than they would like to and need to in order to earn a reasonable living.
Population policy	Encompasses all of the measures taken by a government aimed at influencing population size, growth, distribution or composition.
Pro-natalist policies	Such policies promote larger families.
Anti-natalist policies	Such policies aim to reduce population growth.

The rapid increase in the world's population PAGES 2–4

During most of the early period in which humankind first evolved, global population was very low. Ten thousand years ago, when people first began to domesticate animals and cultivate crops, world population was no more than 5 million. The world's population reached 500 million by about 1650. From this time population grew at an increasing rate. By 1800 global population had doubled to reach 1 billion. Figure 1.1 shows the time taken for each subsequent billion to be reached, with the global total reaching 7 billion in 2011. China and India together account for 36.5 per cent of the world's population.

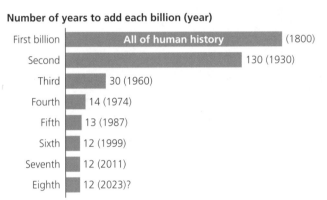

▲ **Figure 1.1** World population growth by each billion

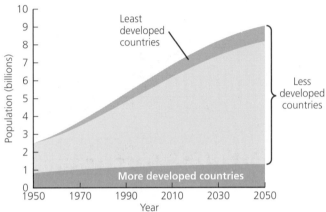

▲ **Figure 1.2** Population growth in more and less developed countries, 1950–2050

Recent demographic change

In 2016, world population increased by 89.8 million, the result of 147.2 million births and 57.4 million deaths. The bulk of this population increase is in the developing countries (Figure 1.2). The very rapid growth of the world's population over the last 70 years or so is the result of the largest ever difference between the number of births and deaths in the world as a whole.

However, only since the Second World War has population growth in the poor countries overtaken that in the rich. The rich countries had their period of high population growth in the nineteenth and early twentieth centuries. For the less developed countries high population growth has occurred since about 1950.

The highest-ever global population growth rate was reached in the early to mid 1960s. At this time the term **population explosion** was widely used to describe this rapid population growth. But by the late 1990s the rate of global population growth was down to 1.8 per cent and by 2016 it had reduced further to 1.2 per cent. However, even though the rate of growth has been falling for about 50 years the number of people added each year remains very high. This is because there are currently so many women in the child-bearing age range.

> **Tip**
>
> It is important to remember that while the world's population continues to increase, the rate of global population growth has been falling for over 50 years.

The causes of a change in population size PAGES 4–5

Population change in a country is affected by (a) the difference between the **birth rate** and the **death rate** (the **rate of natural change**), and (b) the balance between immigration and emigration (**net migration**). For most countries natural change is a more important factor in population change than net migration.

▼ **Table 1.1** Birth and death rates, 2016

Region	Birth rate	Death rate
World	20	8
More developed world	11	10
Less developed world	22	7
Africa	36	10
Asia	18	7
Latin America/Caribbean	17	6
North America	12	8
Oceania	17	7
Europe	11	11

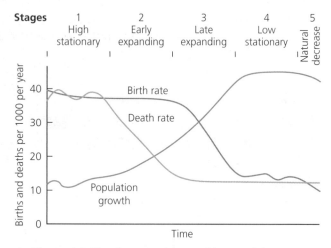

▲ **Figure 1.3** The demographic transition model

The demographic transition model PAGES 5–6

The **demographic transition model** helps to explain the causes of a change in population size (Figure 1.3). No country as a whole retains the characteristics of stage 1, which only applies to the most remote societies on Earth. All the developed countries of the world are now in stage 4 or stage 5. The poorest of the developing countries are in stage 2. Most developing countries which have undergone significant social and economic advances are in stage 3 while some of the first newly industrialised countries such as South Korea and Taiwan have entered stage 4. Stage 5, natural decrease, is mainly confined to Eastern and Southern Europe at present.

- **The high stationary stage (stage 1):** The birth rate is high and stable while the death rate is high and fluctuating due to the sporadic incidence of famine, disease and war. Population growth is very slow and there may be periods of decline.
- **The early expanding stage (stage 2):** The death rate declines to levels never before experienced. The birth rate remains at its previous level as the social norms governing fertility take time to change. The rate of natural change increases to a peak at the end of this stage.
- **The late expanding stage (stage 3):** After a period of time social norms adjust to the lower level of mortality and the birth rate begins to decline.
- **The low stationary stage (stage 4):** Both birth and death rates are low. The former is generally slightly higher, fluctuating somewhat due to changing economic conditions. Population growth is slow.
- **The natural decrease stage (stage 5):** In a limited but increasing number of countries, mainly European, the birth rate has fallen below the death rate.

Tip

Population data change frequently over time, so when you quote data you should also state the year to which they apply. For example, in Table 1.1, the birth rate for 'Latin America/Caribbean' in 2012 was 19/1000, as stated in the previous edition of this book.

Contrasts in demographic transition

There are a number of differences in the way that developing countries have undergone population change compared with the experiences of most developed nations before them. In the developing world:

- birth rates in stages 1 and 2 were generally higher
- the death rate fell much more steeply
- some countries had much larger base populations. Thus the impact of high growth in stage 2 and the early part of stage 3 has been far greater
- for those countries in stage 3 the fall in fertility has also been steeper
- the relationship between population change and economic development has been much weaker.

> **Tip**
>
> It is very important to be clear where the boundary lines are between each stage of the demographic transition model and to understand, and be able to explain, why they are in those particular places.

Test yourself

1 Define *rate of natural change*.
2 Which world region has the highest birth rate?
3 When is the world's population projected to reach 8 billion?

Answers on page 125

Reasons for contrasting rates of population change 📖 PAGES 6–12

REVISED ☐

Population change is governed by three factors: fertility, mortality and migration.

Factors affecting fertility

The most common measure of fertility is the birth rate. However, other more detailed measures are also used, such as the **total fertility rate**. Less than 10 countries in the world now have fertility rates over 6.0 while the 10 lowest fertility countries have a total fertility rate of 1.3 or lower.

Common error

REVISED ☐

Error	Why it is wrong
'Birth rate is the most accurate measure of fertility.'	It is only a very broad indicator as it does not take into account the age and sex distribution of a population. The total fertility rate takes into account these factors and is thus a much more accurate measure of fertility.

The factors affecting fertility can be grouped into four categories (Table 1.2).

▼ **Table 1.2** The factors affecting fertility

Demographic	Other population factors, particularly **infant mortality rates**, influence fertility.
Social/cultural	In some societies, particularly in Africa, tradition demands high rates of reproduction. Education, especially female literacy, is the key to lower fertility. In some countries religion is an important factor influencing fertility.
Economic	In many of the least developed countries children are seen as an economic asset. In the more developed world the general perception is reversed and the cost of the child dependency years is a major factor in the decision to begin or extend a family.
Political	There are many examples in the past century of governments attempting to change the rate of population growth for economic and strategic reasons.

Factors affecting mortality

In 1900 the world average for **life expectancy** was about 30 years. It is presently 72 years. The highest life expectancy of 79 years is in North America, while the lowest of 61 years is in Africa.

The causes of death vary significantly between the developed and developing worlds (Figure 1.4). Apart from the challenges of the physical environment in many developing countries, a number of social and economic factors contribute to the high rates of infectious diseases. These include:

- poverty
- poor access to healthcare
- antibiotic resistance
- evolving human migration patterns
- new infectious agents.

Increasing mortality due to HIV/AIDS

Although, in general, mortality continues to fall around the world, in some countries it is rising, due mainly to HIV/AIDS. However, globally, deaths from AIDS are falling. In 2015, 1.1 million people died from AIDS-related causes worldwide – 45 per cent fewer deaths than in 2005. Eastern and southern Africa remained the region most affected. Factors linked to such a high incidence include:

- high levels of other sexually transmitted infections
- the low status of women
- sexual violence
- high mobility, which is mainly linked to migratory labour systems
- ineffective leadership during critical periods in the epidemic's spread.

> **Tip**
>
> Using 'categories' to structure your explanation, as in Table 1.2, can help to produce a logical sequence of arguments for questions requiring detailed answers.

What are the main differences between rich and poor countries with respect to causes of death?

Online Q&A
30 April 2012

Q: What are the main differences between rich and poor countries with respect to causes of death?

A: In high-income countries almost 50% of the deaths are among adults 80 and over. The leading causes of death are chronic diseases: cardiovascular disease, chronic obstructive lung disease, cancers, diabetes or dementia. Lung infection remains the only leading infectious cause of death.

In middle-income countries, chronic diseases are the major killers, just as they are in high-income countries. Unlike in high-income countries, however, HIV/AIDS, tuberculosis and road traffic accidents also are leading causes of death.

In low-income countries around 40% of all deaths are among children under the age of 14. Although cardiovascular diseases together represent the leading cause of death in these countries, infectious diseases (above all HIV/AIDS, lower respiratory infection, tuberculosis, diarrhoeal diseases and malaria) together claim more lives. Complications of pregnancy and childbirth together continue to be a leading cause of death, claiming the lives of both infants and mothers.

▲ **Figure 1.4** World Health Organization – What are the main differences between rich and poor countries with respect to causes of death?

▼ **Table 1.3** The impact of HIV/AIDS

Labour supply	The economically active population reduces as more people fall sick and are unable to work.
Dependency ratio	An increasing death rate in the economically active age group increases the dependency ratio.
Family	AIDS is impoverishing entire families, and many children and old people have to take on the role of carers. There are a large number of orphaned children.
Education	With limited investment in education many young people are still unaware about how to avoid the risk of contracting HIV.
Poverty	There is a vicious cycle between HIV/AIDS and poverty.

Case study: Kenya – a country with a high rate of population growth

- Kenya has a high rate of population growth due to high fertility and falling death rates, particularly in infant mortality.

- Although Kenya's total fertility rate is falling, the population is forecast to grow to 65.9 million by 2030. Rapid population increase puts heavy pressure on a country's resources.

- Kenya has a very high youth dependency ratio (Figure 1.5) with over 42 per cent of the population under 15.

- A rapidly growing population results in a lower amount of land per capita available to farmers and their children.

- Young people who cannot find work on the land often migrate to urban areas.

- Youth unemployment is a considerable problem as the rate of population increase is greater than the rate of job creation.

- Although the poverty rate fell from 47 per cent in 2005 to 38 per cent in 2012, Kenya remains among the most unequal countries in Africa.

- While progress has been made in health, education, infrastructure and other aspects of society, a significant proportion of the population continue to live in fragile conditions with sub-standard access to water, sanitation and energy.

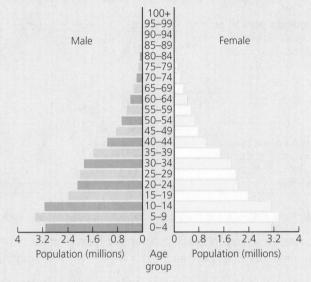

▲ **Figure 1.5** The growth in Kenya's population between 1969 and 2030

Case study: Population decline in Russia

- In 2016, Russia's birth and death rates were equal, at 13/1000. Such stagnant population change or natural decrease is common in Eastern Europe.

- Russia's population reached its highest level of almost 148.7 million in 1991 (Figure 1.6). Since then it has been mainly in decline.

- Population decline/very slow growth has been due to: low birth rates; high death rates, particularly among men; emigration.

- Unemployment and poverty are major concerns for many people. The cost of raising children is perceived to be high when both parents need to work to make ends meet.

- Education standards for women in Russia are high and thus women in general have a major say in decisions about family size. The use of contraception is high.

- In 2016 life expectancy for women was 77 years, but only 66 for men.

- Population decline has had its greatest impact in rural areas, with 8500 villages said to have been abandoned since 2002. The cold northern regions of Russia have experienced the highest levels of **depopulation**.

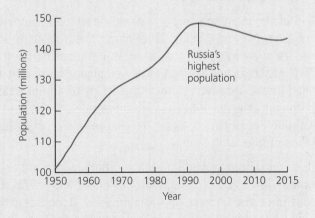

▲ **Figure 1.6** Russia's population 1950–2015

Over-population and under-population

 PAGES 13–15

As a country develops, the highest average living standards mark the **optimum population** (Figure 1.7). Before that population is reached the country or region can be said to be **under-populated**. As the population rises beyond the optimum the country or region can be said to be **over-populated**.

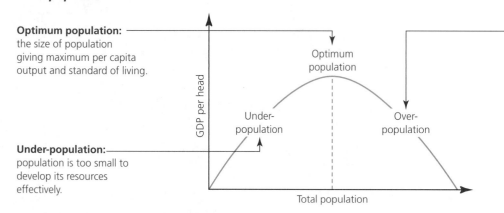

Optimum population: the size of population giving maximum per capita output and standard of living.

Under-population: population is too small to develop its resources effectively.

Over-population: an increase in population or decrease in natural resources which leads to a decrease in standards of living for the population as a whole. This is related to the *carrying capacity* of a country.

▲ **Figure 1.7** Optimum population, over-population and under-population

There are many indications that the human population is pushing up against the limits of the Earth's resources. For example:

- One quarter of the world's children have protein-energy malnutrition.
- The long-term trend for grain production per person is falling.
- Water scarcity already affects every continent and 4 of every 10 people in the world.

The Netherlands and the UK are two of the most densely populated countries in Europe. Signs of population pressure in both countries include:

- intense competition for land
- heavy traffic congestion
- high house prices
- pressure on water resources.

Two of the most sparsely populated developed countries in the world are Australia and Canada. Throughout the history of both countries the general view has been that they would benefit from higher populations. Thus Australia and Canada have welcomed significant numbers of immigrants. However, in recent years, with an uncertain economic climate, both countries have been much more selective in terms of immigration. Although both countries are very large in size, they have large areas of inhospitable landscape.

In the developing world, China and Bangladesh are countries that many would view as over-populated. The 'one-child policy' (changed in 2016) confirmed the Chinese government view. Bangladesh has one of the highest population densities in the world and struggles to provide for many in its population.

Test yourself

4 List the **four** general factors affecting fertility.

5 Define *life expectancy*.

6 Give **two** examples of population pressure.

Answers on page 125

Case study: Bangladesh – an over-populated country?

- At 1128 people per km² Bangladesh's population density is about 20 times the global average. In 1971 the population was about 75 million. By 2016, it had reached 163 million.

- The paucity of natural resources is a major factor in over-population, as is rapid population growth.

- Almost four-fifths of the population live in rural areas. The very small amount of cultivable land per person has resulted in a very high level of rural poverty.

- About 40 per cent of the population is **underemployed**, working a limited number of hours a week on low wages.

- The regular threat of cyclones and flooding hugely exacerbates this problem. Eighty per cent of the country is situated on floodplains.

- Much of the country is close to sea level and about 40 per cent gets flooded during the monsoon season.

- Major floods increase the level of rural to urban migration, with the majority of migrants heading for the capital city Dhaka.

- Living conditions in Dhaka and the other main urban areas are in a very poor state. Many people lack basic amenities. Dhaka has become one of the most crowded cities in the world.

- Poor governance and corruption have hindered development in Bangladesh. However, the number of people in poverty in Bangladesh fell from 63 million in 2000 to 47 million in 2010.

Case study: Australia – an under-populated country?

- Australia is generally regarded as an example of an under-populated country.

- With a population of only 24 million in 2016, the population density is only 3 per km².

- Australia is a resource-rich nation, exporting raw materials all over the world.

- Australia also has great potential for renewable energy, particularly in terms of wind and solar power.

- The country has a well-developed, highly skilled population and generally high incomes. It attracts potential migrants from many countries.

- Australia scores highly for most measures of the quality of life, including health and education.

- Although Australia's population is highly concentrated in certain areas, there are more opportunities for population increase here than in most other parts of the world.

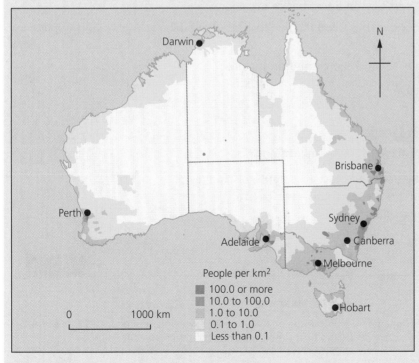

▲ **Figure 1.8** Population density map of Australia

The effectiveness of population policies

 PAGES 16–19

China, with a population in excess of 1.3 billion, operates the world's most severe **anti-natalist policy**. The balance between population and resources has been a major cause of concern for some time. After the communist revolution in 1949 and prior to the 1970s China had periods when it encouraged population growth and when it tried to reduce the population growth rate. The controversial one-child policy was imposed in 1979.

The one-child policy has been most effective in urban areas where the traditional bias of couples wanting a son has been significantly eroded. However, the story is different in rural areas where the strong desire for a male heir remains the norm. In most provincial rural areas, government policy was relaxed so that couples could have two children without penalties.

Although the one-child policy reduced China's birth rate considerably, it caused other problems:

- The policy has had a considerable impact on the sex ratio.
- China had 32 million more men aged under 20 than women.
- China's low birth rate, 12/1000 in 2016, has contributed to the country's ageing population.

In 2016, the government relaxed the rules to allow all couples to have two children.

A small, but growing, number of countries now see their fertility as too low. Some have pursued clear **pro-natalist policies**. Such countries are concerned about:

- the socioeconomic implications of population ageing
- the decrease in the supply of labour
- the long-term prospect of population decline.

France has taken steps to encourage fertility on a number of occasions over the last 80 years, beginning in 1939 when the government passed the 'Code de la Famille'.

Recent measures to encourage couples to have more children include:

- longer maternity and paternity leave
- higher child benefits
- improved tax allowances for larger families
- preferential treatment in the allocation of government housing.

Overall, France is trying to reduce the economic cost to parents of having children. With a total fertility rate of 1.9 (2016), France is close to the replacement level of 2.1 children per woman.

Common error

Error	Why it is wrong
'The one-child policy was the first time China had tried to reduce fertility.'	While the one-child policy was introduced in 1979, this was not the first time China had tried to reduce fertility.

Sample exam question

a Define *birth rate*. [2]

Student's answer

> a   The number of births per 1000 population.

b Describe and explain the rate of natural change at each stage of the model of demographic transition. [5]

Student's answer

> b   In stage 1 the birth rate is high and slightly above the death rate, which varies due to factors such as disease and famine. There is a low rate of natural increase.
>
> In stage 2 the birth rate remains high while the death rate falls significantly. The rate of natural increase becomes greater as stage 2 progresses, reaching a maximum at the boundary between stages 2 and 3.
>
> In stage 3 the birth rate begins to fall, gradually reducing the gap with the death rate. As a result the rate of natural increase declines to reach a low level at the end of this stage.
>
> In stage 4 birth and death rates are low, resulting in a low rate of natural increase.
>
> In stage 5 the birth rate is lower than the death rate, resulting in natural decrease.
>
> In summary, natural increase is low in stages 1 and 4. It is high in stages 2 and 3. Stage 5 is characterised by natural decrease.

Teacher's comments

The student has achieved 1 mark out of the maximum of 2 because no time frame has been stated. The correct definition is 'The number of live births per 1000 population per year'. At IGCSE/GCSE a student would not be penalised for omitting 'live'.

Teacher's comments

This is a very good answer, scoring the maximum 5 marks. The student has accurately described and explained the situation in each of the five stages as well as supplying a concise summary.

Exam-style questions

1 a Define the birth rate and the total fertility rate. [3]
 b Which continent has the highest total fertility rate and which has the lowest? [2]
 c Suggest why the total fertility rate is a more detailed measure of fertility than the birth rate. [3]
2 a How does the infant mortality rate influence fertility? [3]
 b Explain the relationship between education and fertility illustrated by Figure 1.9. [4]

Answers on page 128

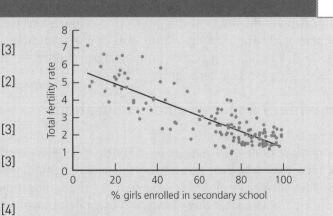

▲ **Figure 1.9** A comparison between female secondary education and total fertility rates

1.2 Migration

Key objectives

You should be able to:

- explain and give reasons for population migration
- demonstrate an understanding of the impacts of migration.

Case study required

- An international migration.

Key definitions

REVISED

Term	Definition
Migration	The movement of people across a specified boundary, national or international, to establish a new permanent place of residence.
Push and pull factors	Push factors are negative conditions at the point of origin, which encourage or force people to move. In contrast, pull factors are positive conditions at the point of destination, which encourage people to migrate.
Refugees	People forced to flee their homes due to human or environmental factors and who cross an international border into another country.
Internally displaced people	People forced to flee their homes due to human or environmental factors who remain in the same country.
Rural-to-urban migration	The movement of significant numbers of people from the countryside to towns and cities.
Remittances	Money sent back to their families in their home communities by migrants.

The nature of and reasons for population migration 📖 PAGES 20–23

REVISED

Migration is the result of the interaction of **push and pull factors** (Figure 1.10). For example, a high level of unemployment is a major push factor in a region or a country. An important pull factor is often much higher wages in another country or region.

Voluntary and involuntary migrations

In voluntary migration the individual has a free choice about whether to migrate or not. In involuntary migration, people are made to move against their will. This may be due to human or environmental factors. The forcible movement of people from parts of the former Yugoslavia under the policy of 'ethnic cleansing' is a recent example of involuntary migration. Migrations may also be forced by natural disasters or by environmental catastrophe.

In recent decades some of the world's worst conflicts have been in the developing world. These troubles have led to numerous population movements on a significant scale. Not all have crossed international frontiers to merit the term **refugee** movements. Instead many are **internally displaced people**. The current conflict in Syria has produced large numbers of both refugees and internally displaced people. Major natural disasters such as the Pakistan floods of 2010 create large numbers of internally displaced people.

The United Nations High Commission for Refugees (UNHCR) put the number of forcibly displaced people worldwide at 65 million at the end of 2015. This included 21.3 million refugees, the remainder being internally displaced people.

> **Tip**
>
> Remember that forced migration is not just the result of armed conflict, but can also occur due to environmental factors such as volcanic eruptions and desertification.

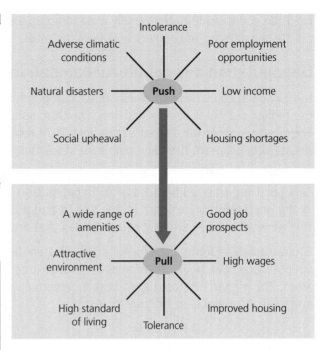

▲ **Figure 1.10** Push and pull factors

Common error

Error	Why it is wrong
'Immigration and emigration have the same meaning.'	Immigration is migration into a country and emigration is migration out of a country.

Migration trends

In 2015, 244 million people lived outside the country of their birth, higher than ever before. This is about 3.3 per cent of the world's population. The number of international migrants more than doubled in the 30 years to 2015. Globalisation has led to an increased awareness of opportunities in other countries. With advances in transportation and communication and a reduction in the real cost of both, the world's population has never had a higher level of potential mobility.

Europe — 76 (2015), 56 (2000)
Asia — 75 (2015), 49 (2000)
North America — 54 (2015), 40 (2000)
Africa — 21 (2015), 15 (2000)
Latin America and the Caribbean — 9 (2015), 7 (2000)
Oceania — 8 (2015), 5 (2000)

Number of migrants (millions)

Key
■ 2015
■ 2000

▲ **Figure 1.11** Bar graph showing number of international migrants by major area of destination, 2000 and 2015

Internal population movements

Population movement within countries is at a much higher level than movements between countries. In both developed and developing countries significant movements of people take place from poorer regions to richer regions as people seek employment and higher standards of living. In developing countries, much of this migration is from rural to urban areas. Developed countries had their period of high **rural-to-urban migration** in the nineteenth century and the early part of the twentieth century. Developing countries have been undergoing high rural-to-urban migration since about 1950, resulting in the very rapid growth of urban areas such as Cairo, Nairobi and Dhaka. The largest rural-to-urban migration in history is now taking place in China where more than 150 million people have moved from the countryside to the rapidly expanding urban/industrial areas to satisfy the demand for workers in China's factories.

Depopulation and counterurbanisation

In developed countries two major trends can be identified concerning the redistribution of population since the late eighteenth century. The first, urbanisation, lasted until about 1970, while the second, counterurbanisation, has been dominant since that time.

The process of urbanisation had a big impact on many rural areas where depopulation occurred because of it. Depopulation is an absolute decline in the population of an area, usually due to a high level of out-migration. It is generally the most isolated rural areas that are affected.

Counterurbanisation is the process of population decentralisation as people move from large urban areas to smaller urban settlements and rural areas. The objective is usually to seek a better quality of life by getting away from the problems of large cities.

Common error

REVISED

Error	Why it is wrong
'Confusing immigration and emigration with in-migration and out-migration.'	Immigration and emigration are the terms used for crossing international borders. In-migration and out-migration are internal movements within one country.

The impacts of migration PAGES 24–28

REVISED

Figure 1.12 shows some of the possible impacts of international migration. Many of these factors are also relevant to internal migration.

The impact of international migration		
Impact on countries of origin	**Impact on countries of destination**	**Impact on migrants themselves**
Positive • Remittances are a major source of income in some countries. • Emigration can ease the levels of unemployment and underemployment. • Reduces pressure on health and education services and on housing.	• Increase in the pool of available labour may reduce the cost of labour to businesses and help reduce inflation. • Increasing cultural diversity can enrich receiving communities. • An influx of young migrants can reduce the rate of population ageing.	• Wages are higher than in the country of origin. • There is a wider choice of job opportunities. • They have the ability to support family members in the country of origin through remittances.
Negative • Loss of young adult workers who may have vital skills, e.g. doctors, nurses, teachers, engineers (the 'brain-drain' effect). • An ageing population in communities with a large outflow of (young) migrants. • Agricultural output may suffer if the labour force falls below a certain level.	• Migrants may be perceived as taking jobs from people in the long-established population. • Increased pressure on housing stock and on services such as health and education. • A significant change in the ethnic balance of a country or region may cause tension.	• The financial cost of migration can be high. • Migration means separation from family and friends in the country of origin. • There may be problems settling into a new culture (assimilation).

▲ **Figure 1.12** Matrix showing the impact of migration

Remittances

Remittances are often seen as the most positive impact on the country of origin. They are a major economic development factor in developing countries. Remittances to developing countries are estimated to have totalled $429 billion in 2016. Remittances exceed considerably the amount of official aid received by developing countries. These revenue flows:

● help alleviate poverty

● spur investment and create a multiplier effect.

The major sources of remittances are the USA, Western Europe and the Persian Gulf. In 2016, the top recipients of remittances were: India, China, the Philippines and Mexico.

Case study: International migration from Mexico to the USA

One of the largest labour migrations in the world has been from Mexico to the USA (Figure 1.13). This is the largest immigrant community in the world. Most migration has taken place in the last four decades. This migration has largely been the result of:

● much higher average incomes in the USA

● lower unemployment rates in the USA

● the faster growth of the labour force in Mexico

● the much better quality of life in the USA.

In the USA the Federation for American Immigration Reform (FAIR) has opposed large-scale immigration from Mexico arguing that it:

● undermines the employment opportunities of low-skilled US workers

● has negative environmental effects because of the increased population

● threatens established US cultural values.

Those opposed to FAIR see its actions as uncharitable and arguably racist. Such individuals and groups highlight the advantages that Mexican and other migrant groups have brought to the country.

The impact of this migration on Mexico includes:

● the high value of remittances, which totalled over $24 billion in 2014

● reduced unemployment pressure as migrants tend to leave areas where unemployment is particularly high

● lower pressure on housing stock and public services

● changes in population structure with emigration of young adults, particularly males

● loss of skilled and enterprising people

● migrants returning to Mexico with changed values and attitudes.

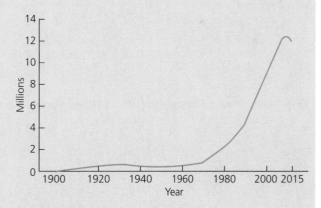

▲ **Figure 1.13** Increase in the Mexican-born population in the USA

Exam-style questions

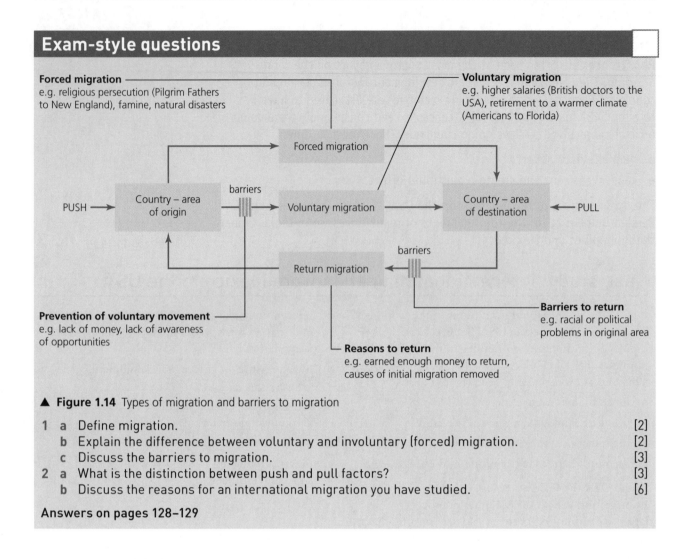

Forced migration
e.g. religious persecution (Pilgrim Fathers to New England), famine, natural disasters

Voluntary migration
e.g. higher salaries (British doctors to the USA), retirement to a warmer climate (Americans to Florida)

Forced migration

PUSH → Country – area of origin — barriers → Voluntary migration → Country – area of destination ← PULL

Return migration — barriers

Prevention of voluntary movement
e.g. lack of money, lack of awareness of opportunities

Barriers to return
e.g. racial or political problems in original area

Reasons to return
e.g. earned enough money to return, causes of initial migration removed

▲ **Figure 1.14** Types of migration and barriers to migration

1 a Define migration. [2]
  b Explain the difference between voluntary and involuntary (forced) migration. [2]
  c Discuss the barriers to migration. [3]
2 a What is the distinction between push and pull factors? [3]
  b Discuss the reasons for an international migration you have studied. [6]

Answers on pages 128–129

Key objectives

You should be able to:

- identify and give reasons for and implications of different types of population structure.

Case study required

- A country with a high dependent population.

Key definitions

Term	Definition
Population structure	The composition of a population, the most important elements of which are age and sex.
Population pyramid	A bar chart arranged vertically, that shows the distribution of a population by age and sex.
Dependency ratio	The ratio of the number of people under 15 and over 64 years to those 15–64 years of age.

Variations in population structure PAGES 29–33

The structure of a population is the result of the processes of fertility, mortality and migration. The most studied aspects of **population structure** are age and sex. Age and sex structure can be illustrated by the use of **population pyramids**. Each bar represents a five-year age group. The male population is represented to the left of the vertical axis with females to the right.

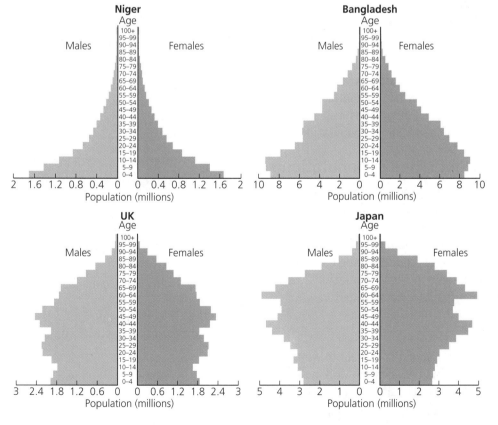

▲ **Figure 1.15** Four population pyramids for Niger, Bangladesh, the UK, Japan

Hodder & Stoughton Limited © Paul Guinness and Garrett Nagle 2019

Demographic transition and changing population structure

Population pyramids change in shape as a country progresses through demographic transition and economic development. In Figure 1.15 the wide base of Niger's pyramid reflects extremely high fertility. The base of the pyramid for Bangladesh is narrower than that of Niger reflecting a considerable fall in fertility after decades of government-promoted birth control programmes. In the pyramid for the UK much lower fertility still is illustrated by narrowing of the base. The final pyramid (Japan) has a distinctly inverted base, reflecting the lowest fertility of all four countries. Table 1.4 shows how demographic data differ between these four countries at different stages of demographic transition.

▼ **Table 1.4** Population data changes with stage of demographic transition

Country	Stage of demographic transition	Birth rate	Death rate	Infant mortality rate	Life expectancy	Pop under 15 (per cent)	Pop 65+ (per cent)
Niger	2	49	9	56	61	50	3
Bangladesh	3	20	5	38	72	33	6
UK	4	12	9	4	81	18	17
Japan	5	8	10	2	84	13	27

Population structure: differences within countries

In countries where there is strong rural-to-urban migration, the population structures of the areas affected can be markedly different. These differences show up clearly on population pyramids. Out-migration from rural areas is age-selective with single young adults and young adults with children dominating this process. Thus the bars for these age groups in rural areas affected by out-migration will indicate fewer people than expected in these age groups.

In contrast, the population pyramids for urban areas attracting migrants will show age-selective in-migration, with substantially more people in these age groups than expected. Such migrations may also be sex-selective. If this is the case it should be apparent on the population pyramids.

Implications of different types of population structure

The fact that Niger and Bangladesh have large numbers of young people in their populations has implications for both countries. For example, these young people have to be housed, fed, educated and looked after in terms of health. All this costs money and governments have to allocate resources to cater for these needs. Conversely, countries such as the UK and Japan have large numbers of older people in their populations. Older people have different needs that governments have to provide for.

The dependency ratio

Dependents are people who are too young or too old to work. The **dependency ratio** is the relationship between the working population and the non-working population. The formula for calculating the dependency ratio is:

$$\text{dependency ratio} = \frac{\text{population aged 0–14 + population aged 65 and over}}{\text{population aged 15–64}} \times 100$$

Tip

When describing and explaining population pyramids a good starting point is to divide the pyramid into three sections: the young dependent population; the economically active population; the elderly dependent population. You can then comment on each section in turn.

Test yourself

1 At what stages of demographic transition are Bangladesh, Japan, Niger and the UK?
2 Define *dependency ratio*.
3 What does a dependency ratio of 80 mean?

Answers on page 125

A dependency ratio of 60 means that for every 100 people in the economically active population there are 60 people dependent on them. The dependency ratio in developed countries is usually between 50 and 75, but in developing countries it may reach over 100. In developing countries, children form the great majority of the dependent population. An increase in the dependency ratio can cause significant financial problems for governments if it does not have the financial reserves to cope with such a change.

Case study: The Gambia – a country with a high dependent population

- The Gambia has a young and fast-growing population. This has placed big demands on the resources of the country.

- 95 per cent of the country's population are Muslim. Until recently religious leaders were against the use of contraception. In addition, cultural tradition meant that women had little influence on family size.

- Children were viewed as an economic asset because of their help with farming. One in three children aged 10–14 is working.

- In 2016 the infant mortality rate was 45/1000. With 46 per cent of the population classed as young dependents and only 2 per cent elderly dependents, the dependency ratio is 92.

- Many parents in the Gambia struggle to provide basic housing for their families. There is huge overcrowding and lack of sanitation, with many children sharing the same bed.

- Rates of unemployment and underemployment are high and wages are low, with parents struggling to provide even the basics for large families.

- Many schools operate a two-shift system with one group of pupils attending in the morning and a different group attending in the afternoon.

- Another sign of population pressure is the large number of trees being chopped down for firewood. As a result desertification is increasing at a rapid rate.

- In recent years the government has introduced a family planning campaign that has been accepted by religious leaders.

Exam-style questions

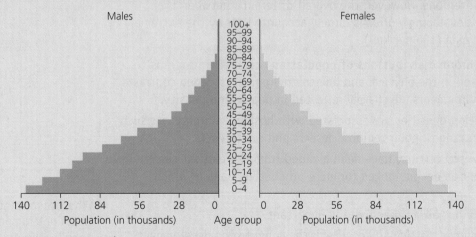

▲ **Figure 1.16** Population pyramid for the Gambia

1 a What aspects of population structure are shown in a population pyramid? [2]
   b On Figure 1.16, draw lines to divide the population pyramid into the young dependent population, the economically active population, and the elderly dependent population. [3]
   c Describe the age structure of the Gambia. [3]

2 a Define the dependency ratio. [2]
   b What does a dependency ratio of 80 mean? [1]
   c How does the structure of dependency vary between developed and developing countries? [4]

Answers on page 129

Key objectives

You should be able to:

- describe the factors influencing the density and distribution of population.

Case studies required

- A densely populated country or area (at any scale from local to regional).
- A sparsely populated country or area (at any scale from local to regional).

Key definitions

REVISED

Term	Definition
Population density	The average number of people per square kilometre in a country or region.
Population distribution	The way that the population is spread out over a given area, from a small region to the Earth as a whole.
Densely populated	Areas with a high population density.
Sparsely populated	Areas with a low population density.

Population density and distribution PAGES 34–37

REVISED

The average **population density** in the less developed world is more than two and a half times that of the more developed world. North America (16 per km²) and Oceania (5 per km²) have the lowest population densities of all the world regions. However, the overall difference between the developed and developing worlds is largely accounted for by the extremely high figure for Asia (136 per km²).

- The most uniform **distributions of population** occur where there is little variation in the physical and human environments. Steep contrasts in these environments are sharply reflected in population density.
- Low population densities are associated with hostile environments such as mountain ranges, polar regions, deserts and rainforests.
- Areas of low soil fertility have been avoided from the earliest times of settlement as people looked for more productive areas in which to settle.
- Water supply has always been vitally important.
- Mineral resources, particularly coalfields, have led to the development of large numbers of settlements in many countries. Although mining may eventually cease when the resource runs out, the investment in infrastructure over time usually means that the settlement will continue.
- The more advanced a country is the more important the elements of human infrastructure become in influencing population density and distribution.

> **Tip**
>
> When describing variations in population density on a map with, say, four colours or types of shading, refer to each class (for example, over 100 per km²) to produce a detailed answer.

Common error

Error	Why it is wrong
'Population density does not change.'	Population density can change considerably over time. Population density is increasing most in regions and countries with the fastest rates of population growth.

Case study: Sparsely and densely populated areas in North America

North America has a low population density compared with most other parts of the world. The USA has an average of 35 people per km², while Canada has only 4 per km². In both countries population is highly concentrated in some areas while large expanses of land elsewhere are very sparsely settled.

The Canadian Northlands: a sparsely populated region

The Canadian Northlands is one of the most sparsely populated regions in the world. Table 1.5 summarises the region's most important characteristics.

▼ **Table 1.5** Characteristics of the Canadian Northlands

Location	Located north of 55 degrees N.
Population density	Less than one person per km².
Temperature	Most of the region has a mean January temperature below –20 °C. Summers are short.
Permafrost	Much of the region is affected by permafrost. Here the ground is permanently frozen to a depth of about 300 metres.
Economic activities	Very limited. Much of the region is beyond the limits of agriculture. Most employment is in the exploitation of raw materials, public services, tourism and defence installations.
Transport	Very limited. Not one of the railway lines in the region crosses the Arctic Circle. The road system is also very sparse. Many water transportation routes are frozen over for much of the year. For many communities air transport is the only link to the outside world.
Settlement	Yellowknife, population 19 000, is the largest settlement in the Northwest Territories.

The northeast of the USA: a densely populated region

- In the USA the most **densely populated** region is in the northeast.
- The region stretches inland from Boston and Washington to Chicago and St Louis.
- The other main cities in this region are New York, Philadelphia and Baltimore.
- New York is a 'global city'. The population of the larger Metropolitan Area of New York is 18.9 million.
- The region also contains many smaller urban areas. Much of the area has an average density over 100 per km².
- The rural parts of the region are generally fertile and intensively farmed. The climate and soils at this latitude are conducive to agriculture. Many people living in the rural communities commute to work in the towns and cities.
- The region has the most highly developed transport networks in North America.

Test yourself

1 Define *population density*.
2 What is the population density of the Canadian Northlands?
3 Name **four** major cities in the northeast of the USA.

Answers on page 125

Exam-style questions

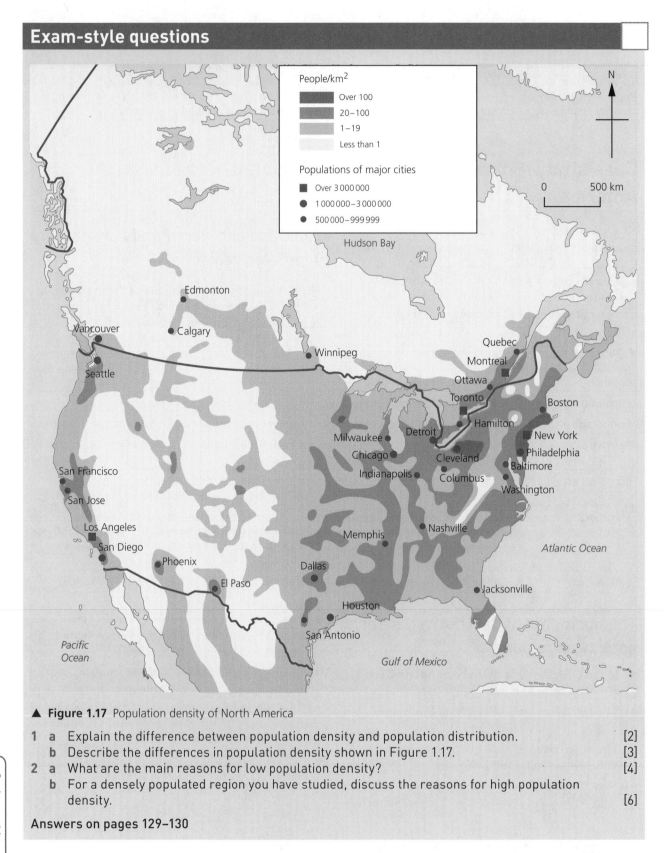

People/km²

▨	Over 100
▨	20–100
▨	1–19
☐	Less than 1

Populations of major cities

■	Over 3 000 000
●	1 000 000–3 000 000
•	500 000–999 999

0   500 km

N

▲ **Figure 1.17** Population density of North America

1 a Explain the difference between population density and population distribution.   [2]
   b Describe the differences in population density shown in Figure 1.17.   [3]
2 a What are the main reasons for low population density?   [4]
   b For a densely populated region you have studied, discuss the reasons for high population
      density.   [6]

Answers on pages 129–130

1.5 Settlements and service provision

Key objectives

You should be able to:

- explain the patterns of settlement
- describe and explain the factors which may influence the sites, growth and functions of settlements
- give reasons for the hierarchy of settlements and services.

Case studies required

- Settlement and service provision in an area.

Key definitions

REVISED ☐

Term	Definition
Dispersed settlement	When farms or houses are set among their fields or spread along roads.
Nucleated settlement	Houses and buildings are tightly clustered around a central feature.
Linear pattern	Settlements are found along a geographical feature such as a river valley or a major transport route.
Site	The actual land on which a settlement is built.
Situation	The relationship between a settlement and its surrounding area.
Low-order functions	Basic functions found in smaller settlements (e.g. hamlets).
High-order functions	More specialised functions and services found in larger settlements (villages and market towns).
Range of a good	The maximum distance a person is prepared to travel to buy a good.
Threshold population	The minimum number of people needed to support a good or service.
Sphere of influence	The area that a settlement serves.

Pattern PAGES 38–40

REVISED ☐

A **dispersed settlement** pattern occurs when farms or houses are set among their fields or spread along roads, rather than concentrated on one point.

Nucleated settlements are ones in which houses and buildings are tightly clustered around a central feature such as a church, village green or crossroads. A number of factors favour nucleation:

- defence, for example hilltop locations or sites within a meander
- shortage of water causing people to locate in areas close to springs
- near important junctions and crossroads for trade and communications.

A **linear pattern** occurs when settlements are found along a geographical feature such as a river valley or a major transport route.

Site and situation PAGE 40

The **site** of a settlement is the actual land on which a settlement is built whereas the **situation** or position is the relationship between a particular settlement and its surrounding area. The advantages and disadvantages of alternative sites for agriculture and housing include:

- availability of water – necessary for drinking, cooking, washing, as a source of food supply and transport
- relief – high enough not to flood and level sites to build on
- fertile soils – often located close to rivers
- good accessibility – increases the potential for trade and commerce, such as close to bridges, weirs, confluence sites, estuaries and at points of navigation
- access to resources such as minerals.

> **Test yourself**
>
> 1 Distinguish between *dispersed* and *nucleated* settlements.
>
> 2 Distinguish between the *site* and *situation* of a settlement.
>
> **Answers on page 125**

Growth of settlements PAGES 40–42

There are a number of factors that affect settlement growth. Areas that are too hot or cold, wet or dry usually have small, isolated settlements. In contrast, in areas where food production is favoured, settlements have managed to grow. Settlements in more favoured areas have greater potential for growth.

Function of settlements PAGES 40–42

Function refers to what a settlement does. Some settlements have a dormitory function. Others may have an agricultural function. Others may be tourist locations, mining villages, fishing villages, and so on.

Settlement hierarchy PAGES 42–44

The term hierarchy means order. Only basic or **low-order functions** are found in the smaller hamlets whereas the same functions and services are found in larger settlements (villages and market towns) together with more specialised ones – **high-order functions**.

The maximum distance that a person is prepared to travel to buy a good is known as the **range of a good**. Low-order goods have a small range whereas high-order goods have a large range. The number of people needed to support a good or service is known as the **threshold population**. Low-order goods may only need a small number of people to support a small shop, whereas high-order goods require a greater threshold population.

> **Test yourself**
>
> 3 State **one** example of a high-order function and **one** example of a low-order function.
>
> 4 Define the terms *threshold population* and *sphere of influence*.
>
> **Answers on page 125**

> **Tip**
>
> When providing examples (for example, low-order goods or high-order goods), give real-life examples or examples from your own area if possible and appropriate, or refer to the examples in the textbook (pages 42–43).

The area that a settlement serves is known as the **sphere of influence**. Hamlets and villages generally have low spheres of influence whereas larger towns and cities have large spheres of influence.

In general, as population size in settlements increases, the number and range of services increase (see the textbook, Figure 1.68, page 44). However, there are exceptions. Some small settlements, notably those with a tourist-related function, may be small in size but have many services. In contrast, some dormitory (commuter) settlements may be quite large but offer few functions or services other than a residential one. In these settlements, people live (reside) in the village but work and shop elsewhere.

Case study: Settlement hierarchy at Lozère, France

Lozère is located in southeast France. It is a mountainous region, and the main economic activities are farming and tourism. Due to the mountainous relief and poor-quality soil, farming is mainly cattle rearing. Surprisingly, the region has a very low rate of unemployment. This is due to a long history of out-migration of young people in search of work. Table 1.12 (page 45 of the textbook) shows how the population of St-André-Capcèze fell between the 1860s and the mid 2000s (Figure 1.18). Nevertheless, in recent years the population has increased slightly due to improved communications and easier travel. However, the population is an ageing one. Tourism offers some employment, and the services available relate partly to tourist potential (see the textbook, Table 1.14, page 45).

▲ **Figure 1.18** Population change in St-André-Capcèze, 1800–2006

Exam-style questions

1 Compare the characteristics of a linear village shape with those of a nucleated village shape. [2]
2 Describe the relationship between population size and number of services, as shown in Figure 1.66, page 43 of the textbook. [2]

Answers on page 130

1.6 Urban settlements

Key objectives

You should be able to:

- describe and give reasons for the characteristics of, and changes in, land use in urban areas
- explain the problems of urban areas, their causes and possible solutions.

Case studies required

- An urban area (including changing land use and urban sprawl).

Key definitions

Term	Definition
Urban land use	Activities such as industry, housing and commerce that may be found in towns and cities.
Bid rent	When land value and rent decrease as distance from the central business district increases.
Central business district	An area of an urban settlement where most of the commercial activity takes place.
Suburbs	The outer part of an urban settlement, generally consisting of residential housing and shops of a low order.
Rural–urban fringe	The boundary of a town or city, where new building is changing land use from rural to urban.
Urban sprawl	Occurs when urban areas continue to grow without any form of planning.
Urban redevelopment	Attempts to improve an urban area.
Urban renewal	When existing buildings are improved.
Gentrification	The movement of higher social or economic groups into an area after it has been renovated and restored.

Characteristics of urban areas PAGES 48–69

Urban land use refers to the activities such as industry, housing and commerce that may be found in towns and cities. The concept of **bid rent** states that land value and rent decrease as distance from the central business district increases (see the textbook, Figure 1.73 page 49).

The **central business district (CBD)** is where most of the commercial activity is found. It is the most accessible (to public transport) and has the highest land values. It tends to have high-rise buildings owing to the high demand for land, but a shortage of space.

Most residential areas are found in the suburbs. The **suburbs** refer to the outer part of an urban area. Suburbs generally consist of residential housing and shops of a low order (newsagent, small supermarket). Often, suburbs are the most recent growth of an urban area. Their growth may result in urban sprawl. In contrast, the **rural–urban fringe** is the boundary of a town or city, where new building is changing land use from rural to urban. It is often a zone of planning conflict.

Industrial areas are found in a number of locations such as the inner city (the area surrounding the CBD), along major transport routes, and in edge-of-town locations. In many cities the inner city is the older industrial area of the city and may suffer from decay and neglect, leading to social problems. Inner cities are characterised by poor-quality terraced housing with old manufacturing industry nearby.

However, urban areas are changing rapidly. Much retailing and commerce is now taking place on the edge of town. Inner city areas are being used for residential purposes. There are also important differences between high-income countries (HICs) and low-income countries (LICs). LICs lack the same industrial development that is associated with HICs. In addition, many LICs have illegal settlements, known as shanty towns. Nevertheless, some shanty towns are changing and becoming more up-market ('gentrified').

Land use zoning in LICs PAGES 50–58

There are a number of models that describe and explain the development of cities in LICs. There are several key points:

- the rich generally live close to the city centre whereas the very poor are more likely to be found on the periphery
- higher quality land is occupied by the wealthy
- segregation by wealth, race and ethnicity is evident
- manufacturing is scattered throughout the city.

Urban change PAGES 54–55, 67–68

Many urban areas prove very attractive for young workers. This is on account of job availability and access to better services. Consequently, some urban areas continue to grow without any form of planning. This is known as **urban sprawl** – the unchecked outward spread of built-up areas caused by their expansion. Urban sprawl may be prevented by the use of green belts – areas surrounding a city in which urban development is severely limited.

Much of the change has been in edge-of-town locations. This is because land prices are lower, land is available for development and accessibility to private cars is high. Edge-of-town sites have become important for retailing, industry and the provision of services.

Some parts of urban areas have gone into decline. **Urban redevelopment** attempts to improve an urban area, where existing buildings are either demolished and rebuilt or renovated. Other forms of improvement include **urban renewal**, whereby existing buildings are improved. Both of these methods may be carried out by the government or a mix of government and private developers.

In contrast, **gentrification** is the movement of higher social or economic groups into an area after it has been renovated and restored. This may result in the out-migration of the people who previously occupied the area. It most commonly occurs in the inner city. This was originally a feature of HICs but has been seen in some LICs such as in Woodstock in Cape Town (South Africa) and Vidigal in Rio de Janeiro (Brazil).

> **Tip**
>
> A model is a simplification. You should not expect any city to illustrate all of the characteristics of any one model, although they may show some of them.

> **Test yourself**
>
> 1 Distinguish between *urban sprawl* and *urban renewal*.
> 2 Explain the meaning of the term *gentrification*.
>
> **Answers on page 125**

Common error

Error	Why it is wrong
'Confusing the CBD with the inner city.'	The inner city is (or in some cases, was) the industrial area surrounding the CBD. If you use the term (former) industrial area it should help avoid confusion. However, there are other industrial areas such as ports, along major transport routes, and on the edge of cities.

Sample exam question

1 Explain two characteristics of the CBD. [4]

Student's answer

The CBD has lots of shops and offices. It also has many high-rise buildings.

Teacher's comments

This is purely descriptive and does not explain the characteristics. The CBD has many shops and offices because it is very accessible and can be reached by many potential customers and workers. The buildings are high-rise because there is a shortage of land, and the land value is very high. Therefore, developers create new land by building upwards. Two marks awarded.

Case study: Urban sprawl

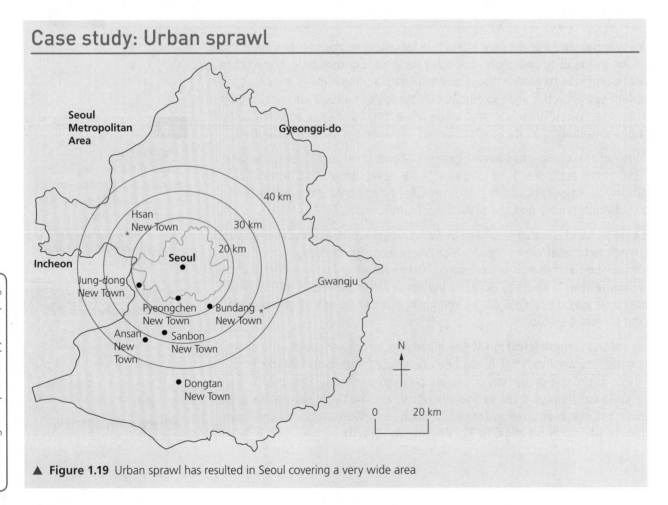

▲ **Figure 1.19** Urban sprawl has resulted in Seoul covering a very wide area

Seoul has grown dramatically since the early 1960s. It currently has a population of about 10 million in the main city (municipality) and between 26 million and 36 million in the Metropolitan Region (Figure 1.19). This includes important cities in their own right, such as Incheon and Gwanju, as well as new towns. Seoul experiences a number of problems, just like most large cities. These include pollution, inequality, housing, traffic congestion and conflicts over land use change.

Problems in Seoul

Pollution

As Seoul has grown, the amount of air and water pollution has increased. A good example is the Cheong Gye Cheon River in central Seoul. It had become heavily polluted with lead, chromium and manganese and was a health risk. It was even covered up by an elevated highway. Restoration of the river has been a central part of the regeneration of central Seoul. Previously up to 87 per cent of the city's sewage flowed untreated into the Hangang River. Now Seoul has the capacity to treat up to 3 million tonnes of sewage each day.

Inequality

There has been increased inequality in Seoul since the financial crisis of 1997. Unemployment has increased and there is a greater gap between rich and poor in Seoul, compared with any other city in Korea. Many migrants travel to Seoul every year in the hope of a good job. However, not all of these migrants will get good jobs. The richer area of Seoul is Gangnam-Gu, to the south of the city. This is a relatively new high-tech industrial area with good accessibility, good schools and high-income residents. In contrast, the poor tend to be north of the river (or in the far west) in areas of traditional manufacturing.

Housing shortage

Seoul's population has grown from 2.5 million in 1960 to around 10 million today and over 25 million in the Seoul Metropolitan Region. In-migration and the trend towards nuclear families (two generations) rather than the extended family (three generations in the one house) have created a major housing shortage, despite massive building programmes. Less than 45 per cent of the land around Seoul is available for urban development due to steep terrain and mountains. The type of housing is changing too. The typical one-storey one-family house with inner courtyards is being replaced by high-rise apartment blocks. Such flats have increased from 4 per cent of housing in 1970 to 35 per cent in 1990 and 50 per cent today. Until recently most of the housing was to the north of the river, but a number of satellite towns have been built to the south of the river. This has evened out population density, which is, on average, over 16 000 people per km².

Traffic congestion

Seoul experiences massive traffic congestion. In 1975 South Korea manufactured fewer than 20 000 cars. By 1994 there were over 2 million cars registered in the Seoul area. Despite improvements to the motorway network, the increase in the population of Seoul and the number of cars in the area mean that congestion has increased. In addition, many of the roads in central Seoul are relatively small and unable to handle the large volumes of traffic.

Land use conflicts

There are many pressures on land for development in the Seoul region. In 1971, Seoul introduced a green belt system (Restricted Development Zone). Many people were against this as it lowered the value of their house (and could not be improved by developers). However, in 1979 the government released 40 per cent of the green belt land for development. Opponents state that these new developments will add to the urban sprawl, will increase over-crowding and that there should be conservation of natural resources, such as open spaces and forests/parks, for future generations. Many of the new developments are taking place towards the edge of the city or indeed in other cities at some distance from Seoul (see Figure 1.19 for the distribution of new towns around Seoul) and Figure 1.87 on page 63 of the textbook to see the distribution of green belt land in Seoul.

Exam-style questions

1   Study Figure 1.107 photo (a) on page 78 of the textbook. Suggest the environmental problems likely to be experienced in the area shown in the photo.   [4]
2   a   Study Figure 1.95 on page 68 of the textbook. Describe the changes in Detroit's population between 1900 and 2015.   [3]
    b   Suggest reasons for its changes.   [4]

Answers on page 130

Key objectives

You should be able to:

- identify and suggest reasons for rapid urban growth
- describe the impacts of urban growth on both rural and urban areas, along with possible solutions to reduce the negative impacts.

Case studies required

- A rapidly growing urban area in a developing country and migration to it.

Key definitions

REVISED

Term	Definition
Urbanisation	The process by which the proportion of a population living in or around towns and cities increases through migration and natural increase.
Millionaire city	A city with over 1 million inhabitants.
Megacity	A city with over 10 million inhabitants.

Rapid urban growth PAGES 70–71

REVISED

Urbanisation is the process by which the proportion of a population living in or around towns and cities increases through migration and natural increase. It is one of the most important geographical features of the twentieth and twenty-first centuries. Not only are more people living in urban areas (over 50 per cent of the world's population now live in urban areas), but people are also living in larger settlements. Figure 1.20

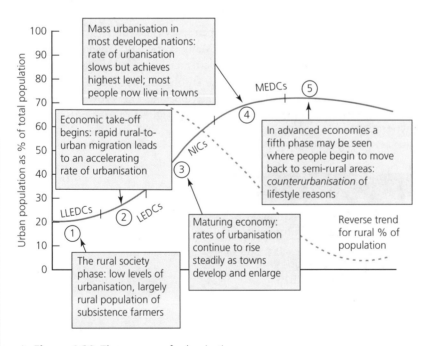

▲ **Figure 1.20** The process of urbanisation

shows that the most rapid urban growth is taking place in LICs and newly industrialising countries (NICs). In contrast, in many HICs, the urban population is declining, due to an increase in counterurbanisation and the movement into rural areas.

There are a number of reasons for rapid urbanisation in LICs and NICs. Some of these can be referred to as urban pull factors and rural push factors. These include:

- the prospects of finding employment, better paid jobs and more secure jobs in urban areas (economic pull factor)
- better provision of education and health facilities in urban areas (social pull factor)
- fewer economic opportunities in rural areas – farming can be low paid, insecure and subject to climate and natural hazards (economic and physical push factors)
- poor access to clean water and sanitation, healthcare and education in rural areas (social push factors).

Consequently, rural–urban migration may be large-scale, leading to rural depopulation and the growth of population in urban areas. The majority of those who migrate are young adults. Consequently, the rural area has a population that is ageing, whereas the urban area has a youthful population. The rural area therefore has higher death rates and lower birth rates on account of its ageing population. In contrast, the younger population in the urban areas has higher birth rates and lower death rates, contributing to the increase of the urban population.

The effects of urbanisation on the people and natural environment

In addition to the changes to the population structure, rural–urban migration has a lasting effect on the rural economy. The departure of many young adults removes some of the workforce and many skilled workers. In extreme cases, most of the men will leave and women are left to raise their families and look after the farm. Productivity may be reduced and poverty levels increase. Some farmland may even be abandoned.

In rural areas close to the urban area, land may be used for development or for squatter settlements. There may be theft of farm produce.

The characteristics of squatter settlements

Squatter settlements are illegal settlements, in which the residents have no legal rights. Often, the buildings may be made from any available materials (for example, wood, corrugated iron, textiles and cardboard). Squatter settlements generally lack access to running water and proper sanitation, as well as electricity and gas supplies. Housing density is generally high, with many people sharing a room. However, in some places, squatter settlements are accepted and residents may upgrade their housing. A distinction is made between 'slums of hope', which are located close to jobs, and 'slums of despair', which are located further away from economic activity.

Strategies to reduce the negative impacts of urbanisation 📖 PAGES 77–83

There are a number of options to reduce the negative impacts of urbanisation:

● create more jobs in rural areas so that people do not need to leave the area

● legalise squatter settlements and give residents security

● provide more running water and sanitation both in rural and urban areas

● provide loans and subsidies for people to improve their homes.

Case study: Urbanisation in China

Since economic reforms began in China in 1978, between 150 million and 200 million Chinese have migrated from rural to urban areas.

Shanghai

Shanghai has a population of over 24 million.

Shanghai experiences a number of problems such as rapid in-migration, housing shortages, income inequality, pollution, congestion and poor air quality.

Housing and demographic issues

Housing shortages and overcrowding problems are acute. Almost half the population lives in less than 5 per cent of the total land area, and in central Shanghai population density reaches 40 000–160 000 people per km². Population pressure is caused by in-migration, overcrowding, disparities in wealth and the social insecurity of Shanghai's poor 'floating population', i.e. the migrant labourers. From the 1990s whole neighbourhoods were demolished. Over 2 million residents were moved to the outer suburbs to live in better quality accommodation.

Economic growth has attracted an increasing number of foreign migrants and Chinese living overseas to live in Shanghai. Many of these live in luxury, gated apartments. This increases the rich–poor gap in Shanghai.

Water and air quality

Water quality in Shanghai is a concern: less than 60 per cent of waste water and storm water and less than 40 per cent of sewage flows are intercepted, treated and disposed of. Waste disposal is also a major problem: the Huangpu River receives 4 million cubic metres of untreated human waste every day.

▲ **Figure 1.21** Special economic zones and open cities

Potential solutions

To deal with population issues, the Shanghai government has a series of important policies:

- a combination of widespread family planning and medical care, to reduce fertility levels among the young immigrant population
- compulsory work permits to control the number of migrants.

Water and waste

Since the 1990s there have been major improvements in access to water and waste disposal. Most houses now have piped water, electricity and waste disposal. Organic waste is now used as fertiliser in surrounding rural areas.

Transport

The Shanghai authorities have invested greatly in transport. Eight tunnels and four bridges have been built over/under the Huangpu River. Shanghai's underground system, with a daily capacity of 1.4 million, is now linked to Pudong airport by the world's fastest commercial magnetic levitation train – MAGLEV – capable of reaching 431 km per hour. Other strategies to improve safety have been pedestrianisation and a reduction in the number of bicycles, currently estimated at 9 million and a cause of many road accidents.

Shanghai master plan

In 2000 Shanghai introduced the New Master Plan for Shanghai (2000–20). This includes the whole area and the development of three satellite cities (new towns) to accommodate Shanghai's growth. The aim is to reduce congestion and high population densities in central Shanghai.

Test yourself

1 Define the term *urbanisation*.
2 Distinguish between *megacities* and *millionaire cities*.

Answers on page 125

Common errors

REVISED

Error	Why it is wrong
'Giving the answer "e.g. Africa" or "e.g. Asia" when asked for a named city in a country.'	You must have a named country and, preferably, a named city within that country.
'Urbanisation is rural-to-urban migration.'	Urbanisation includes rural-to-urban migration, but it is also caused by natural increase. It is defined as the increase in the proportion of people living in urban areas.

Tip

When asked for an example of a squatter settlement, many students write 'Rio' or 'Cairo', for example. Neither are squatter settlements. Rocinha and Vidigal are squatter settlements in Rio de Janeiro, just as the City of the Dead is a squatter settlement in Cairo.

Exam-style questions

1 Suggest reasons for rapid urbanisation in low-income countries. [4]
2 Contrast the advantages and disadvantages of living in squatter settlements. [5]
3 Outline ways in which it is possible to improve housing in squatter settlements in LICs. [3]

Answers on page 130

Key objectives

You should be able to:

- describe the main types and features of volcanoes and earthquakes
- describe and explain the distribution of earthquakes and volcanoes
- describe the causes of earthquakes and volcanic eruptions and their effects on people and the environment

- demonstrate an understanding that volcanoes present hazards and offer opportunities for people
- explain what can be done to reduce the impacts of earthquakes and volcanoes.

Case studies required

- An earthquake.
- A volcano.

Key definitions

REVISED

Term	Definition
Crater	Depression at the top of a volcano following a volcanic eruption. It may contain a lake.
Lava	Molten magma that has reached the Earth's surface. It may be liquid or may have solidified.
Shield volcano	Gently sloping volcano produced by very hot, runny lava.
Cone volcano	Steeply sloping volcano produced by thick lava.
Ash	Very fine-grained volcanic material.
Cinders	Small-sized rocks and coarse volcanic materials.
Magma	Molten rock within the Earth. When magma reaches the surface it is called lava.
Magma chamber	The reservoir of magma located deep inside the volcano.
Pyroclastic flow	Superhot (700 °C) flows of ash, pumice (volcanic rocks) and steam at speeds of over 500 km per hour.
Vent	The channel through which volcanic material is ejected.
Dormant	Volcanoes which have not erupted for a very long time but could erupt again.
Active	A volcano currently showing signs of activity.
Extinct	A volcano which has shown no signs of volcanic activity in historic times.
Intensity	The power of an earthquake is generally measured using the **Richter scale** or sometimes the **Mercalli scale**.
Richter scale	An open-ended scale to record magnitude of earthquakes – the higher the number on the scale, the greater the strength of the earthquake. There are more small earthquakes than large earthquakes.
Mercalli scale	Relates ground movement to commonplace observations of, for example, light bulbs, book cases and building damage.
Epicentre	The point on the Earth's surface directly above the focus of an earthquake. The strength of the shock waves generally decrease away from the epicentre.
Focus	The position within the Earth where an earthquake occurs. Earthquakes may be divided into shallow-focus and deep-focus earthquakes depending on how far below the Earth's surface they occur.

Types of volcano PAGES 90–91

The shape of a volcano depends on the type of **lava** it contains. Very hot, runny lava produces gently sloping **shield volcanoes**, while thick material produces **cone volcanoes**.

Composite volcanoes (strato-volcanoes)

These are steep volcanoes formed of sticky (viscous) acidic lava, **ash** and **cinders**. They are found at destructive plate margins. Mt St Helens in the USA is a good example. They may have several secondary cones rather than just a single cone.

Shield volcanoes

Shield volcanoes are low-angle volcanoes formed of runny basaltic lava. They are found at constructive plate margins and hotspots. Mauna Loa in Hawaii is a good example.

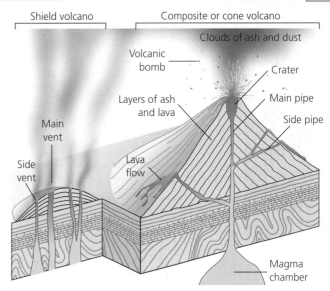

▲ **Figure 2.1** Types of volcano

Features of earthquakes PAGE 91

An earthquake is a sudden, violent movement of the Earth. It occurs after a build-up of pressure causes rocks to give way.

Tectonic plates PAGES 93–95

> **Tip**
>
> The **Richter scale** is logarithmic so an earthquake measuring 7.0 on the Richter scale is 10 times more powerful than one measuring 6.0, and 100 times more powerful than one measuring 5.0.

The global pattern of plates and plate boundaries

The Earth's surface consists of a number of tectonic plates. There are seven major plates: Eurasian, North American, South American, African, Indo-Australian, Pacific and Antarctic. There are many minor plates such as the Caribbean, Juan de Fuca, Cocos, Aegean, Adriatic and Turkish.

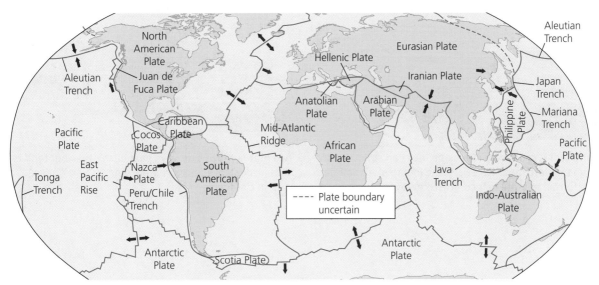

▲ **Figure 2.2** The world's main tectonic plates

Structure of plates

The crust is the outer layer of the Earth's surface. There are two main types of crust: continental crust and oceanic crust. Their main differences are summarised in Table 2.1.

▼ **Table 2.1** A comparison of oceanic crust and continental crust

Type of plate	Continental crust	Oceanic crust
Thickness	35–70 km on average	6–10 km on average
Age of rocks	Very old, mainly over 1500 million years	Young, mainly under 200 million years
Appearance of rocks	Lighter, with an average density of 2.6; light in colour	Heavier, with an average density of 3.0; dark in colour
Nature of rocks	Numerous types, many contain silica and oxygen; granite is the most common	Few types, mainly basalt

> **Tip**
>
> Although the map of plate boundaries is well known, in reality plate boundaries are often not clear-cut, and there are many areas where the plate boundaries are uncertain. Scientists do not know everything about the restless Earth.

Processes at plate boundaries

▼ **Table 2.2** Types of plate margin

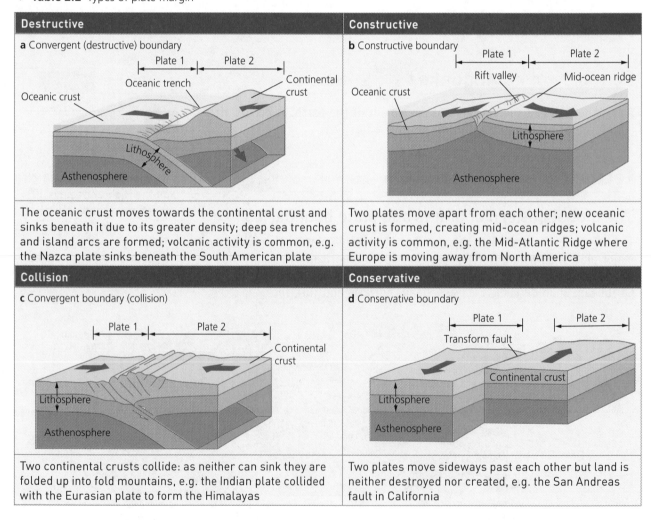

Destructive	Constructive
a Convergent (destructive) boundary	**b** Constructive boundary
The oceanic crust moves towards the continental crust and sinks beneath it due to its greater density; deep sea trenches and island arcs are formed; volcanic activity is common, e.g. the Nazca plate sinks beneath the South American plate	Two plates move apart from each other; new oceanic crust is formed, creating mid-ocean ridges; volcanic activity is common, e.g. the Mid-Atlantic Ridge where Europe is moving away from North America
Collision	**Conservative**
c Convergent boundary (collision)	**d** Conservative boundary
Two continental crusts collide: as neither can sink they are folded up into fold mountains, e.g. the Indian plate collided with the Eurasian plate to form the Himalayas	Two plates move sideways past each other but land is neither destroyed nor created, e.g. the San Andreas fault in California

Test yourself

1 State the difference between *dormant* and *extinct* volcanoes.
2 Distinguish between *destructive* and *constructive* plate boundaries.

Answers on page 125

Causes of earthquakes and volcanoes PAGES 96–97 REVISED

Most earthquakes occur at plate boundaries. They are caused by the release of pressure as two plates move in different directions or at different speeds. Other earthquakes may be caused by human activity such as:

- nuclear testing
- the weight of large dams
- drilling for oil/fracking.

Volcanoes also mainly occur at plate boundaries. However, some occur in the middle of a plate, at locations known as hotspots. Over a long period of time, rising **magma** enters into a **magma chamber**. When the pressure in the chamber is great enough, an eruption may occur.

> **Tip**
>
> Earthquakes may occur anywhere: some of the largest ones in the USA have been at great distances from plate boundaries. This makes them difficult – if not impossible – to predict with accuracy (for example, where, when, how big?). Volcanic eruptions are also difficult to predict (how big and when?).

Test yourself

3 Explain how humans can cause earthquakes.

Answer on page 125

Impacts of earthquakes and volcanic eruptions on people and the environment PAGES 97–106 REVISED

Earthquakes

▼ **Table 2.3** Earthquake hazards and impacts

Primary hazard	Secondary hazard	Impacts
• Ground shaking • Surface faulting	• Ground failure and soil liquefaction • Landslides and rockfalls • Debris flows and mudflows • Tsunamis	• Loss of life • Loss of livelihood • Total or partial destruction of building structure • Interruption of water supplies • Breakage of sewage disposal systems • Loss of public utilities such as electricity or gas • Floods from collapsed dams • Release of hazardous material • Fires • Spread of chronic illness

Volcanoes

▼ **Table 2.4** Hazards associated with volcanic activity

Direct hazards	Indirect hazards	Socio-economic impacts
Pyroclastic flows • Volcanic bombs (projectiles) • Lava flows • Ash fallout • Volcanic gases • Lahars (mudflows) • Earthquakes	• Atmospheric ash fallout • Landslides • Tsunamis • Acid rainfall	• Destruction of settlements • Loss of life • Loss of farmland and forests • Destruction of infrastructure – roads, airstrips and port facilities • Disruption of communications • Reduced tourist arrivals • Lack of investment • Fewer jobs • Reduced earnings for farmers • Decreased productivity • Out-migration

Opportunities provided by volcanoes

Volcanoes provide many opportunities for human activities:

- new land and islands for people to live on
- fertile soils
- soils rich in minerals
- important as tourist destinations.

On the other hand they can kill people and destroy properties and livelihoods.

Case study: Soufrière Hills, Montserrat

The Soufrière Hills volcano erupted in 1995 after being **dormant** for nearly 400 years. Its biggest eruption was in 1997 when 19 people were killed. The volcano is caused by the subduction of the Atlantic plate under the Caribbean plate.

To reduce the impact of the volcano, the capital city Plymouth was evacuated, and most of the islanders fled to the north of the island or overseas. Emergency shelters and facilities were provided in the north of the island. In the long term there has been major redevelopment of housing, schools, hospitals and a new airport.

The Montserrat Volcano Observatory was set up to monitor changes in the volcano. Scientists regularly:

- check the size and shape of the volcano
- use seismometers to check for internal changes within the Earth's crust
- measure emissions of sulfur.

It is possible, in some cases, to divert lava flow by spraying vast volumes of water onto the advancing lava flow. Alternatively, diversion channels have been dug to divert lava flows away from settlements.

Reducing the impact of earthquakes

The main ways of dealing with earthquakes include:

- better forecasting and warning
- building location (see the textbook, Figure 2.16, page 101)
- building design (see the textbook, Figure 2.14, page 100)
- emergency procedures.

There are a number of ways of predicting and monitoring earthquakes. These include measurement and observation of:

- small-scale ground surface changes
- ground tilt
- changes in rock stress
- micro-earthquake activity (clusters of small quakes)
- changes in radon gas concentration
- unusual animal behaviour, especially toads.

Case study: Nepal, 2015 – an earthquake in an LEDC

The 2015 Nepal earthquake – magnitude 7.8 – occurred on 25 April as a result of the Indian plate colliding with the Eurasian plate. The epicentre of the earthquake was 80 km north of the capital, Kathmandu, with a shallow focus of 15 km. It occurred around midday. There were over 300 aftershocks, some reaching magnitudes of over 7.0. Over 9000 people died in the main earthquake and at least 200 were killed in an aftershock of 7.3 in May 2015. Fatalities were much lower in rural parts of Nepal because most people were outdoors, and were not affected by collapsing buildings. The economic cost of the earthquake is estimated at about US$7 billion – around 35 per cent of Nepal's GDP. The Asian Development Bank provided a US$3 million grant for emergency relief, including temporary shelters, food, blankets and cooking utensils.

Exam-style questions

1 State the main differences between cone volcanoes and shield volcanoes. [4]
2 State the meaning of the terms epicentre and focus. [2]
3 Describe the advantages of volcanoes. [3]
4 Compare the primary and secondary effects of earthquakes. [4]

Answers on page 130

2.2 Rivers

Key objectives

You should be able to:

- explain the main hydrological characteristics and processes which operate within rivers and drainage basins
- demonstrate an understanding of the work of a river in eroding, transporting and depositing
- describe and explain the formation of the landforms associated with these processes

- demonstrate an understanding that rivers present hazards and offer opportunities for people
- explain what can be done to manage the impacts of river flooding.

Case studies required

- The opportunities presented by a river, the hazards associated with it and their management.

Key definitions

REVISED

Term	Definition
Tributary	A stream or river which joins a larger river.
Drainage basin	The area of land drained by a river system (a river and its tributaries).
Watershed	A ridge or other line of separation between two river systems.
Confluence	The point at which two rivers meet.
Interception	The precipitation that is collected and stored by vegetation.
Infiltration	The movement of water into the soil. The rate at which water enters the soil (the infiltration rate) depends on the intensity of rainfall, the permeability of the soil, and the extent to which it is already saturated with water.
Throughflow	The downslope movement of water in the subsoil.
Evaporation	The process in which a liquid turns to a vapour.
Overland flow	Overland movement of water after a rainfall. It is the fastest way in which water reaches a river. The amount of overland runoff increases with heavy and prolonged rainfall, steep gradients, lack of vegetation cover, and saturated or frozen soil.
Abrasion (or corrasion)	The wearing away of the bed and bank by the load carried by a river.
Attrition	The wearing away of the load carried by a river. It creates smaller, rounder particles.
Hydraulic action	The force of air and water on the sides of rivers and in cracks.
Groundwater flow	The movement of water from land to river through rock. It is the slowest form of such water movement, and accounts for the constant flow of water in rivers during times of low rainfall.
Suspension	Small particles are held up by turbulent flow in the river.
Saltation	Heavier particles are bounced or bumped along the bed of the river.
Solution	The removal of chemical ions, especially calcium, which cause rocks to dissolve. The chemical load is carried dissolved in the water.
Traction	The heaviest material is dragged or rolled along the bed of the river.

Channel characteristics and processes  PAGES 107–109

There are many changes to rivers as they move from their source (start) to the mouth (end). These changes are summarised in Figure 2.3. Most of the changes that occur result from the contribution of more water from **tributaries** further down the **drainage basin**.

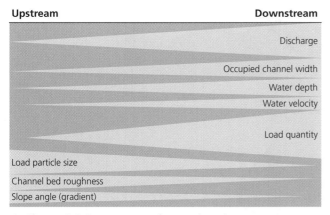

▲ **Figure 2.3** Downstream changes in a river

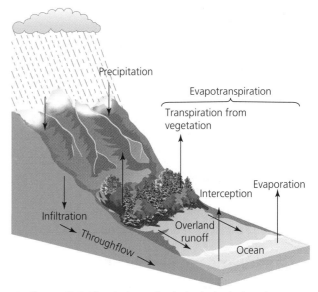

▲ **Figure 2.4** The drainage basin hydrological cycle

The hydrological cycle (water cycle) is the movement of water between air, land and sea (Figure 2.4). It varies in scale from the global hydrological model to a small-scale drainage basin hydrological model. Precipitation includes rain, snow, frost and dew. Evapotranspiration includes the combined losses of water from the ground/water bodies and vegetation.

River processes PAGES 113–120

Factors affecting erosion

The heavier and sharper the load the greater is the potential for erosion. The greater the velocity and discharge the greater is the potential for erosion. Increased gradient increases the rate of erosion. Soft rocks, such as sand and gravel, are more easily eroded. Rates of solution are increased when the water is more acidic. Human impact (for example, deforestation, dams and bridges) interferes with the natural flow of a river and frequently results in an increase in the rate of erosion.

Transport

The main types of transport include **suspension**, **saltation**, **solution**, **traction** and flotation.

Deposition

Deposition occurs due to a decline in energy or velocity. This may happen because a river becomes shallower – for example, when a flood spills onto a floodplain; the river gets trapped behind a dam or enters a lake or the sea; vegetation slows down the river; or the gradient becomes gentler.

> **Tip**
>
> Remember that the factors affecting erosion interact with each other. In any single case, the impact of one factor may be altered through the impact of others.

> **Test yourself**
>
> 1 Compare *hydraulic action* with *abrasion*.
> 2 Outline the difference between *saltation* and *traction*.
>
> Answers on page 125

Common error

Error	Why it is wrong
'Erosion only occurs in the upper course and deposition in the lower course.'	Both processes occur throughout the river's course.

Long and cross-profiles

The shape of a river valley generally changes downstream from being steep and V-shaped in its upper course, to being much flatter and wider in its lower course. The shape of the valley cross-section is known as the cross-profile. The change in gradient from source to mouth is known as the long profile.

There are a number of reasons for these changes including tributary streams, changes in climate and vegetation, human impacts and river processes. The main river processes include erosion, transport and deposition.

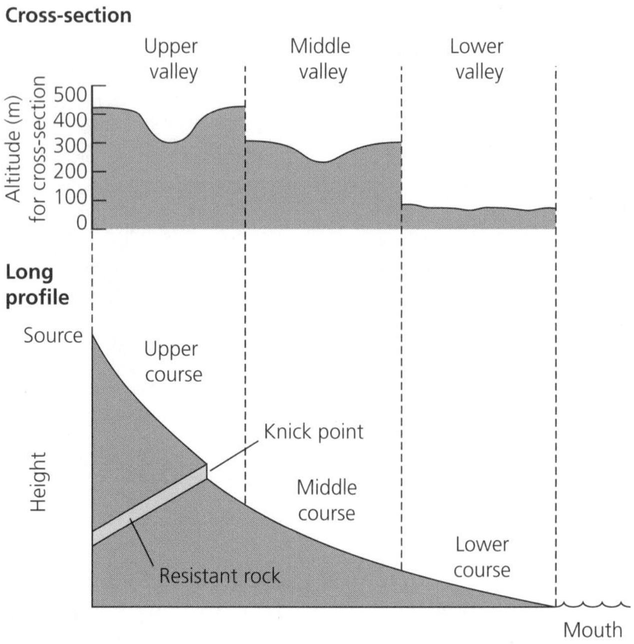

Landforms caused by erosion

These include waterfalls, gorges and potholes. Rivers erode softer rocks and undercut harder rocks. This causes the harder rocks to collapse and form a waterfall (for example, Niagara Falls). If a waterfall retreats, it may form a gorge of recession.

Landforms caused by erosion and deposition

These include meanders and oxbow lakes. Erosion occurs on the outer bank of a meander whereas deposition occurs on the inner bank. This produces a river cliff on the outer bank and a slip-off slope (river beach) on the inner bank. In some cases, the meanders may become so exaggerated that successive meanders touch. In times of flood, the river may break through the meander to form an oxbow lake or meander cut-off.

Landforms caused by deposition

Levées and floodplains are formed by deposition. They occur in the lower course of a river. When the river floods, the larger, coarser material is deposited next to the river channel, whereas the finer clays and silts are carried further away from the river. The coarser sand and gravel may build up a small bar by the edge of the river – a levée – whereas the finer material produces a flat floodplain. When a river enters a lake or the sea where there is very little current, the river slows down and deposits its load, forming a delta.

▲ **Figure 2.5** Long and cross-profiles

Tip

When drawing a diagram of oxbow lakes make sure you label where the erosion and deposition are occurring.

Sample exam question

a Briefly explain how waterfalls and gorges are formed. [4]

Student's answer

a Waterfalls frequently occur on horizontally bedded rocks. The soft rock is undercut by hydraulic action and abrasion, to form a plunge pool. The softer rock is eroded by fragments of the harder rock that break off. The weight of the water and the lack of support cause the waterfall to collapse and retreat. Over thousands of years the waterfall may retreat enough to form a gorge of recession.

b Draw an annotated diagram to show the formation of an oxbow lake. [4]

Student's answer

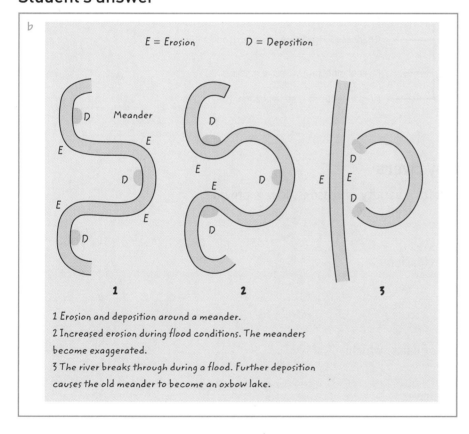

1 Erosion and deposition around a meander.
2 Increased erosion during flood conditions. The meanders become exaggerated.
3 The river breaks through during a flood. Further deposition causes the old meander to become an oxbow lake.

Teacher's comments

The question did not ask for an annotated diagram, although one could have been provided. Some aspects of physical geography are easier to revise and explain using diagrams. Nevertheless, the student has identified: differences in rock strength; the horizontal layering of rocks; the types of erosion; an implied reason for the effectiveness of erosion (hard rock eroding softer rock); progression over time, leading to the formation of a gorge. Full marks awarded.

Teacher's comments

Very clear answer. Shows progression/sequencing. Clearly annotated. Full marks.

River hazards and opportunities  PAGES 121–128

Rivers can cause problems through flooding and river erosion. The flood hazard is extremely dangerous for people's lives and their possessions. Many settlements are built on raised ground in order to reduce the risk of flooding. In addition, rivers may erode their banks thereby making some people extremely vulnerable to losing their homes and fields.

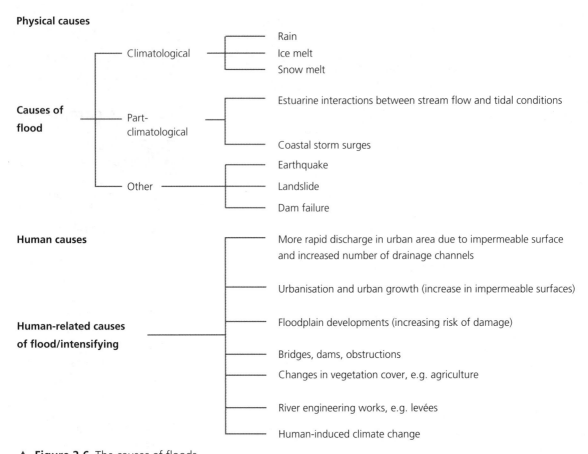

Physical causes

▲ **Figure 2.6** The causes of floods

Opportunities presented by rivers

Rivers are very attractive to people for a variety of reasons. Rivers provide:

- a source of drinking water
- fertile silt for agriculture
- a line of communications and navigation
- a source of power
- fishing
- recreation.

Rivers are affected by the building of dams, which:

- reduce the speed of water flow
- control the amount of water in a river
- cause deposition behind the dam
- increase the amount of erosion downstream of the dam
- change ecosystems
- increase pressure on rocks and may cause earthquakes.

But dams provide:

- reliable water throughout the year
- navigation
- hydro-electric power
- water for irrigation
- safety from flooding.

Test yourself

3 Briefly explain the main causes of floods.

Answer on page 125

Managing the impacts of river flooding

There are many ways in which people can manage the impacts of flooding. These include:

- building dams or reservoirs to hold back excess water (see the case study 'Hard engineering – the Three Gorges Dam' on page 123 of the textbook for more information)
- raising the banks of rivers
- dredging the river channel so that it can hold more water
- diverting streams and creating new flood relief channels
- using sandbags to prevent water getting into houses
- building houses on stilts so that water can pass underneath
- land use planning (i.e. build only on land that is free from flooding)
- afforestation (plant forests) to increase interception and reduce overland flow
- have insurance cover for vulnerable areas and communities
- improved forecasting and warning
- river restoration and allowing rivers to flood naturally (see the case study 'Soft engineering – the Kissimmee River Restoration Project', on pages 124–5 of the textbook for more information).

Even with all these measures flooding cannot be prevented. However, the impacts can be minimised.

Test yourself

4 Briefly explain how flood risk can be managed.

Answer on pages 125–126

Sample exam question

REVISED ☐

1 Outline the hazards and opportunities of living in a named river valley. [7]

Student's answer

The Nile Delta is one of the oldest intensively cultivated areas in the world. It is heavily populated and has a population density of about 16,000 people per km². Only 2.5 per cent of Egypt's land area is suitable for intensive agriculture – up to 95 per cent of Egypt's agricultural production comes from the Nile valley and delta. The delta has long been a source of freshwater and fertile silt, as well as an excellent location for the import and export of goods. The flat land makes building easy. However, it is increasingly under stress.

The delta covers around 25,000 km², is home to around 66 per cent of the country's rapidly growing population and provides over 60 per cent of the nation's food supply. However, most of the delta is very low lying, and an increase in sea level of just 1 m would flood 20 per cent of the area. Flooding by the river Nile is a potential problem. Excessive irrigation has led to waterlogging, whilst significant amounts of fertilisers and pesticides are leached into water courses along the delta. Seawater intrusion has led to the salinisation of groundwater.

Teacher's comments

The student has offered a range of benefits, two of which are supported with quantification. The student has made some general points about the risk of flooding – a recent example would be useful, or mention of the 2016 floods that killed 98 people in the upper Nile valley. A reason for the increased flood risk is given – numbers of people at risk, or names of cities at risk, or dates of the floods would have made this answer complete. Six marks awarded out of a maximum of seven.

Exam-style questions

☐

1 Compare infiltration and throughflow. [2]
2 Compare the cross-section of a river in its upper course with that of a river in its lower course. [4]
3 Compare the long profile of an upper-course river with that of a lower-course river. [2]

Answers on page 130

It is illegal to photocopy this page

2.3 Coasts

Key objectives

You should be able to:

- demonstrate an understanding of the work of the sea and wind in eroding, transporting and depositing
- describe and explain the formation of the landforms associated with these processes
- describe coral reefs and mangrove swamps and the conditions required for their development
- demonstrate an understanding that coasts present hazards and offer opportunities for people
- explain what can be done to manage the impacts of coastal erosion.

Case studies required

- The opportunities presented by an area of coastline, the hazards associated with it and their management.

Key definitions

REVISED

Term	Definition
Abrasion (or corrasion)	The wearing away of the cliffs by the load carried by the sea.
Hydraulic action	The force of air and water when the waves break.
Solution (or corrosion)	The removal of chemical ions, especially calcium, which cause rocks to dissolve.
Attrition	The wearing away of the load carried by the sea.
Fringing reefs	Reefs that grow outwards around an island.
Barrier reef	A reef that is separated from the coast by a deep channel.
Atoll	A circular reef enclosing a shallow lagoon.

Marine processes PAGES 129–131

REVISED

There are a number of processes that occur in coastal zones. These include:

- wave action from constructive and destructive waves
- wind action
- mass movements and weathering
- river and ice actions.

Types of wave

▼ **Table 2.5** Destructive and constructive waves

Destructive waves (erosional waves)	Constructive waves (depositional waves)
Short wavelength	Long wavelength
High height (> 1 metre)	Low height (< 1 metre)
High frequency (10–12/minute)	Low frequency (6–8/minute)
Backwash → swash	Swash → backwash

Processes of transportation

In the water, particles are moved in different ways:

- larger particles are dragged along the sea floor by traction
- smaller particles may be bounced along the sea floor by saltation
- very fine materials are held up in suspension
- dissolved sediments (for example, calcium) may be carried in **solution**.

Deposition

Deposition occurs for a variety of reasons:

- a decrease in wave energy or velocity
- a large supply of material
- an irregular, indented coastline (for example, river mouths).

Wave refraction and longshore drift

Wave refraction occurs when waves approach an irregular coastline or at an oblique angle. Refraction reduces wave velocity and, if complete, causes wave fronts to break parallel to the shore. Wave refraction concentrates energy on the flanks of headlands and disperses energy in bays. However, refraction is rarely complete and consequently longshore drift occurs. The swash is the movement up the beach while the backwash is the movement down the beach.

> **Test yourself**
>
> 1 Distinguish between *destructive* and *constructive* waves.
>
> **Answer on page 126**

Landscapes of erosion 📖 PAGES 131–133   REVISED ☐

Headlands and bays form when rocks are of differing strengths. Hard rocks form resistant headlands while the softer rocks are eroded to form bays.

Cliffs are steep slopes. They occur due to erosion by waves at their base. They may be helped by weathering and mass movements.

Wave-cut platforms are typically less than 500 m wide with an angle of about 1°. Steep cliffs are eroded and replaced by a lengthening platform and lower-angle cliffs, which are then subjected to weathering and mass movements rather than marine forces (Figure 2.7).

On a coastline of a single rock type, weaknesses such as fault lines may be eroded to form a cave. Over time, the cave may be eroded so much that it goes all the way through the rock to form an arch. In time, the arch may be eroded to form a single column, known as a stack.

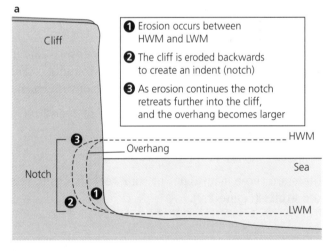

> **Test yourself**
>
> 2 Briefly explain how a wave-cut platform is formed.
>
> **Answer on page 126**

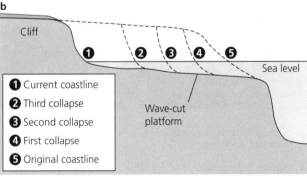

▲ **Figure 2.7** Formation of a wave-cut platform

Features of deposition 📖 PAGES 133–139

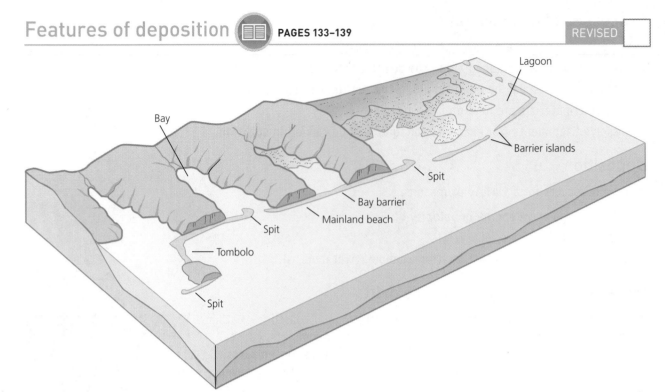

▲ **Figure 2.8** Features of coastal deposition

A beach is a deposit of sand or shingle formed in an area where there is a large supply of material, constructive waves and/or strong onshore winds that carry sediment inwards during low tides.

A spit is a beach of sand or shingle linked at one end to land. They are found on indented coastlines or at river mouths (Figure 2.8).

Coral reefs

Corals are small marine organisms. They absorb calcium salts from seawater and combine them with carbon dioxide to build a skeleton of calcium carbonate. Coral requires certain conditions in which to thrive:

● clear, salt water with a temperature of over 20 °C

● shallow coastal water

● a good supply of water and plankton.

There are three main types of coral reef: **fringing reefs**, **barrier reefs** and **atolls** (Figure 2.9).

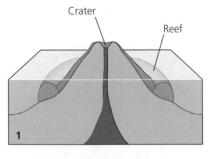

Rocky volcanic islet encircled by fringing coral reef

Reef enlarges as land sinks (or sea rises)

Circular coral reef or atoll (with further change in level)

▲ **Figure 2.9** Formation of coral reefs

Mangroves

Mangroves are salt-tolerant forests of trees and shrubs that grow in the tidal estuaries and coastal zones of tropical areas. The muddy waters are rich in nutrients from decaying leaves and wood. Mangroves cover about 25 per cent of the tropical coastline, the largest being the 570 000 ha mangrove forest in the Sundarbans in Bangladesh.

Coastal hazards and opportunities PAGES 139–143   REVISED

Approximately 40–45 per cent of the world's population live within 150 km of the coast. Coastal areas are popular because they enable trade and commerce; they offer opportunities for tourism and recreation; they allow fishing and energy developments (for example, offshore wind farms). However, coastal areas are vulnerable to coastal erosion, rising sea levels, tropical storms and associated storm surges, and, in some locations, tsunami events.

Tropical storms

Hurricanes are intense tropical storms that bring heavy rainfall, strong winds and high waves, and cause other hazards such as flooding and mudslides. Their paths are erratic, which means it is not always possible to give more than 12 hours' advance warning.

Hurricanes typically:

- develop as intense low-pressure systems over tropical oceans
- have a calm central area, known as the eye, around which winds spiral rapidly
- have a diameter of up to 800 km, though the strong winds that cause most of the damage are found in a narrower belt up to 300 km wide.

In order for a hurricane to develop, water temperatures need to be above 27 °C for sufficient evaporation to occur. Hurricanes also occur away from the Equator, since there is insufficient rotation there. Once they pass over land, they are starved of moisture and begin to lose strength.

The impact of a hurricane depends very much on the degree to which an area is prepared. In general, HICs have better access to satellite equipment (improved monitoring), enforced building codes resulting in stronger buildings (better protection), better communications (to issue warnings, i.e. better prediction), more advanced emergency services, more widespread insurance cover, and more training and evacuation drills (better preparation). Nevertheless, poorer communities in a wealthy country may suffer disproportionately compared with their richer neighbours (for example, the black population in New Orleans as a result of Hurricane Katrina).

Tsunami

The term 'tsunami' is Japanese for harbour wave. Most (90 per cent) occur in the Pacific Basin. The cause of the 2004 South Asian tsunami was a giant earthquake (9.0 on the Richter scale) and landslide caused by the sinking of the Indian plate under the Eurasian plate. Over 240 000 people were killed, mainly in Sumatra, Indonesia. Many homes were wiped out, boats washed away, forest damaged and infrastructure destroyed.

Rising sea levels

Many countries worldwide are under threat from rising sea levels. This is a longer-term consequence of global climate change. Some of the areas most at risk are LICs, such as Bangladesh, although low-lying HICs are also at risk. HICs, however, may have the resources to combat the perceived risk.

Coastal management **PAGES 144–145**   REVISED

Coastal defence covers protection against coastal erosion (coast protection) and flooding by the sea. Coastal zone management is concerned with the whole range of activities that take place in the coastal zone and promotes integrated planning to manage them.

Hard engineering structures

Hard engineering structures include groynes, sea walls, revetments, rock armour and cliff drains (see Table 2.6). They try to alter natural processes to reduce the potential for erosion of the coastline. Sometimes they may have unexpected results (for example, the building of groynes may reduce erosion in nearby locations but increase coastal erosion downdrift). Sea walls may also lead to the scouring of the sea bed.

Soft engineering

Soft engineering refers to working with nature. Examples include the maintenance of a mangrove forest to reduce the impact of tropical storms. Beach nourishment increases the size of a beach by using sediment dredged from elsewhere. Some soft engineering may allow the coastline to retreat naturally.

Managed retreat allows nature to take its course – erosion in some areas, deposition in others. Benefits include less money being spent and the creation of natural environments. However, some homes or farms may be lost to the power of the sea.

Test yourself

4 Distinguish between *gabions* and *sea walls*.

Answer on page 126

▼ **Table 2.6** Different forms of coastal management

Type of management	Aims/methods	Strengths	Weaknesses
Hard engineering	To control natural processes		
Cliff base management	To stop cliff or beach erosion		
Sea walls	Large-scale concrete curved walls designed to reflect wave energy	Easily made; good in areas of high density	Expensive; lifespan about 30–40 years; foundations may be undermined
Revetments	Porous design to absorb wave energy	Easily made; cheaper than sea walls	Lifespan limited
Gabions	Rocks held in wire cages absorb wave energy	Cheaper than sea walls and revetments	Small scale
Groynes	To prevent longshore drift	Relatively low cost; easily repaired	Cause erosion on downdrift side; interrupt sediment flow
Rock armour	Large rocks at base of cliff to absorb wave energy	Cheap	Unattractive; small-scale; may be removed in heavy storms
Offshore breakwaters	Reduce wave power offshore	Cheap to build	Disrupt local ecology
Rock strongpoints	To reduce longshore drift	Relatively low cost; easily repaired	Disrupt longshore drift; erosion downdrift

Type of management	Aims/methods	Strengths	Weaknesses
Cliff face strategies	To reduce the impacts of sub-aerial processes		
Cliff drainage	Removal of water from rocks in the cliff	Cost-effective	Drains may become new lines of weakness; dry cliffs may produce rockfalls
Cliff regrading	Lowering of slope angle to make cliff safer	Useful on clay (most other measures are not)	Uses large amounts of land – impractical in heavily populated areas
Soft engineering	Working with nature		
Offshore reefs	Waste materials, e.g. old tyres, weighted down, to reduce speed of incoming wave	Low technology and relatively cost-effective	Long-term impacts unknown
Beach nourishment	Sand pumped from sea bed to replace eroded sand	Looks natural	Expensive; short-term solution
Managed retreat	Coastline allowed to retreat in certain places	Cost-effective; maintains a natural coastline	Unpopular; political implications
'Do nothing'	Accept that nature will win	Cost-effective!	Unpopular; political implications

Common error

REVISED

Error	Why it is wrong
'Management schemes guarantee safety.'	The 10 m sea walls along the Japanese coastline were not high enough to protect against the 11 m waves generated by the 2011 tsunami.

Tip

Many stretches of coastline have a range of management types – usually they will be a mix of hard and soft engineering, often side by side.

Case study: Opportunities, associated hazards and the management of the Dubai coastline

Some coastal areas provide an excellent opportunity for tourism. Coastal areas that can guarantee hot, sunny conditions have an added advantage. Coastal reclamation in the United Arab Emirates has been developing on a large scale since 2001 (Figure 2.10). Two palm-shaped artificial islands, Palm Jumeirah and Palm Jebel Ali, were completed in 2007. In 2003 plans were unveiled for a third, Palm Deira, as well as 'The World', a collection of over 300 islands, each one in the shape of a country.

Palm Jumeirah not only created a new shoreline; it also became the centre for world-class hotels, over 200 shopping outlets, and a range of luxury housing and leisure and entertainment developments. Sea-front projects ranging from desalination plants to artificial islands have transformed the entire coastline in the past few decades. More than 40 per cent of the shores of some countries in the region are now developed. According to a report in the journal *Nature*, uncontrolled development and a lack of scientific monitoring are seriously threatening ecosystems along this coast.

To create the islands for Palm Jumeirah, some 94 million m³ of sediment were dredged from the sea. Such large-scale projects are changing the ecology in ways that will only become clear in the coming decades. One of the problems is water circulation. Water around some parts of the islands can remain almost stationary for several weeks, increasing the risk of algal blooms. In addition, the fish that have colonised the new environment are invasive species (species from outside the area).

The Gulf region has already lost 70 per cent of its coral reefs since 2001, and most of the remaining reefs are threatened or degraded. Construction of Palm Jebel Ali – a larger archipelago than Palm Jumeirah – has already destroyed 8 km² of natural reef.

Between 2009 and 2015 there was an increase in the size of a number of beaches, for example, Jumeirah One Beach, and the construction of a number of breakwaters in order to protect the coast from medium-height waves. Sand barriers have also been constructed to preserve the coast by preventing coastal erosion and sediment transport. The Dubai Coastal Zone Monitoring Programme records wave height, speed, frequency and direction, water temperature, salinity, dissolved oxygen and sediment movement in the coastal zone.

▲ **Figure 2.10** Coastal engineering schemes in Dubai

Exam-style questions

1 Describe the process of longshore drift. [3]
2 Explain how a stack is formed. [4]
3 Explain the formation of spits. [4]
4 Identify the hazards associated with tropical storms (hurricanes). [2]
5 Suggest why the Nile Delta is vulnerable to sea level change. [3]
6 Identify the advantages of the Nile Delta for people. [4]

Answers on page 131

2.4 Weather

Key objectives

You should be able to:

- describe how weather data are collected
- make calculations using information from weather instruments
- use and interpret graphs and other diagrams showing weather and climate data.

Key definition

REVISED

Term	Definition
Isohyet	A line on a map which joins areas of equal rainfall.

Measuring the weather PAGES 149–153

REVISED

Stevenson screen

A Stevenson screen is a wooden box standing on four legs at a height of about 120 cm (Figure 2.11). The screen is raised so that air temperature can be measured. The sides of the box are slatted to allow air to enter freely. The roof is usually made of double boarding to prevent the Sun's heat from reaching the inside of the screen. Insulation is further improved by painting the outside of the screen white to reflect much of the Sun's energy. The screen is usually placed on a grass-covered surface, thereby reducing the radiation of heat from the ground.

Instruments kept inside the Stevenson screen include a maximum-minimum thermometer and a wet- and dry-bulb thermometer (also called a hygrometer).

Rain gauge

A rain gauge is used to measure rainfall. It consists of a cylindrical container, in which there is a collecting can containing a glass or plastic jar, and a funnel that fits on to the top of the container (Figure 2.12). It is important to check the rain gauge every day, preferably at the same time.

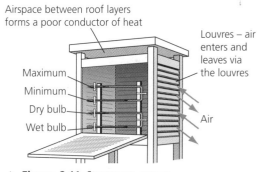

Airspace between roof layers forms a poor conductor of heat

Louvres – air enters and leaves via the louvres

Maximum
Minimum
Dry bulb
Wet bulb

Air

▲ **Figure 2.11** Stevenson screen

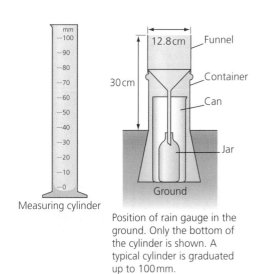

mm
—100
—90
—80
—70
—60
—50
—40
—30
—20
—10
—0

Measuring cylinder

12.8 cm — Funnel

30 cm

Container

Can

Jar

Ground

Position of rain gauge in the ground. Only the bottom of the cylinder is shown. A typical cylinder is graduated up to 100 mm.

▲ **Figure 2.12** Rain gauge

Maximum-minimum thermometer

Maximum thermometer

When the temperature rises, the mercury expands and pushes the index along the tube (Figure 2.13). When the temperature falls, the mercury contracts and the index remains behind. The maximum temperature is obtained by reading the scale at the base of the index, which was in contact with the mercury. The index is then drawn back to the mercury by a magnet.

Minimum thermometer

When the temperature falls, the alcohol contracts and its meniscus pulls the index along the tube. When the temperature rises, the alcohol expands. The daily readings of maximum-minimum thermometers are used to work out the average or mean temperature for one day (called the mean daily temperature) and the temperature range for one day (called the daily or diurnal temperature range).

To find the mean daily temperature, the maximum and minimum temperatures for one day are added together and then halved. For example: (maximum temperature 35 °C + minimum temperature 25 °C) ÷ 2 = mean daily temperature 30 °C.

Wet- and dry-bulb thermometer (hygrometer)

Wet- and dry-bulb thermometers are used to measure relative humidity. The dry-bulb is a glass thermometer which records the actual air temperature. The wet-bulb is a similar thermometer, but with the bulb enclosed in a muslin bag which dips into a bottle of water (Figure 2.14). This thermometer measures the wet-bulb temperature which, unless the relative humidity is close to 100 per cent, is generally lower than the dry-bulb temperature.

Sunshine recorder

The number of hours and minutes of sunshine received at a place can be measured and recorded by a sunshine recorder (such as a Campbell-Stokes sunshine recorder – Figure 2.15).

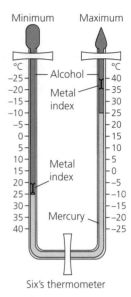

Six's thermometer

▲ **Figure 2.13** Maximum-minimum thermometer

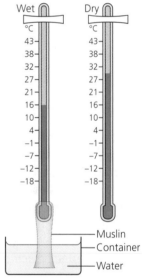

▲ **Figure 2.14** Wet- and dry-bulb thermometer

Test yourself

1 Outline the main features of a *Stevenson screen*.

Answer on page 126

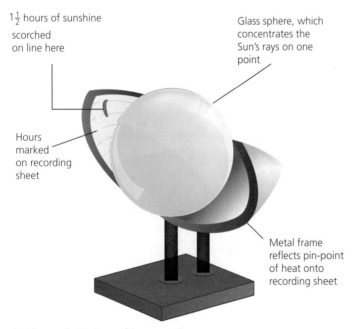

$1\frac{1}{2}$ hours of sunshine scorched on line here

Glass sphere, which concentrates the Sun's rays on one point

Hours marked on recording sheet

Metal frame reflects pin-point of heat onto recording sheet

▲ **Figure 2.15** A sunshine recorder

Barometer

A mercury barometer consists of a hollow tube from which the air is extracted before the open end is placed in a bath of mercury. Mercury is forced up the tube by the pressure of the atmosphere on the mercury in the bath (Figure 2.16). When the pressure of the mercury in the tube balances the pressure of the air on the exposed mercury, the mercury in the tube stops rising. The height of the column of mercury changes as air pressure changes (i.e. it rises when air pressure increases and falls when air pressure decreases).

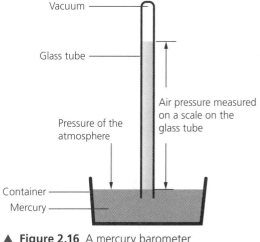

▲ **Figure 2.16** A mercury barometer

Anemometer

An anemometer is used to measure wind speed. It consists of three or four metal cups fixed to metal arms that rotate freely on a vertical shaft (Figure 2.17). When there is a wind, the cups rotate. The stronger the wind, the faster is the rotation. The number of rotations is recorded on a meter to give the speed of the wind in km/hr.

Wind vane

A wind vane is used to indicate wind direction. It consists of a horizontal rotating arm pivoted on a vertical shaft (Figure 2.18). The rotating arm has a tail at one end and a pointer at the other. When the wind blows, the arm swings until the pointer faces the wind. The directions north, east, south and west are marked on arms that are rigidly fixed to the shaft.

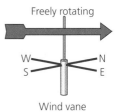

▲ **Figure 2.18**
A wind vane

Digital instruments

Digital instruments can be used for weather observations. The main advantage is that they give a reliable reading whereas other methods are more subjective.

▲ **Figure 2.17** An anemometer

Recording the weather 📖 PAGE 154 REVISED ☐

Observations of types and amounts of cloud

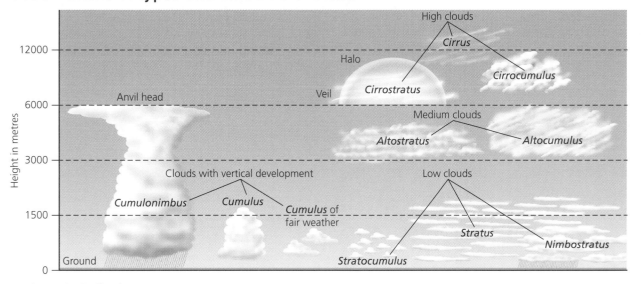

▲ **Figure 2.19** Cloud types

Clouds can be separated into three broad categories according to the height of their base above the ground: high clouds, medium clouds and low clouds (Figure 2.19). High clouds are usually composed solely of ice crystals and have a base between 5500 and 14 000 m. Medium clouds are usually composed of water droplets or a mixture of water droplets and ice crystals, and have a base between 2000 and 7000 m. Low clouds are usually composed of water droplets, though cumulonimbus clouds include ice crystals, and have a base below 2000 m.

Cloud cover

Cloud cover is measured in oktas (eighths). This is made by a visual assessment of how much of the sky is covered by cloud. If the sky is completely covered with clouds it has 8/8 cloud cover.

Climate graphs

Figure 2.20 shows two simple climate graphs (or climographs). Climate graphs tell us a great deal about the pattern of temperature and rainfall. They are often used to show annual variations or sometimes variations over a few weeks.

The mean monthly average temperature occurs between the mean monthly maximum and the mean monthly minimum. (The mean monthly maximum is the average of all the maximum temperatures for each day of the month. The mean monthly minimum is the average of all of the minimum temperatures recorded for each day in a month.)

Rainfall is normally shown as a bar chart. Different scales are used – in this case temperature is shown on the left-hand side and rainfall on the right-hand side.

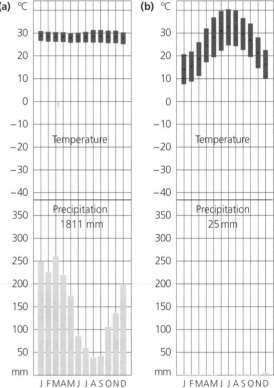

Exam-style questions

Study the climate graphs in Figure 2.20. For both climate graphs: [1]
1 State the mean monthly temperature for July. [1]
2 State the mean monthly temperature for February. [2]
3 Calculate the annual temperature range.
4 Describe the pattern of rainfall over the year. [3]

Answers on page 131

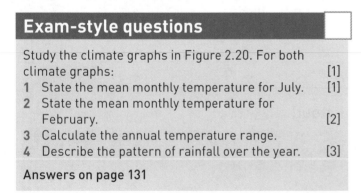

▲ **Figure 2.20** Climate graphs for (a) Manaus and (b) Cairo

Key objectives

You should be able to:

- describe and explain the characteristics of two climates: equatorial and hot desert
- describe and explain the characteristics of tropical rainforest and hot desert ecosystems
- describe the causes and effects of deforestation of tropical rainforest.

Case studies required

- An area of tropical rainforest.
- An area of hot desert.

Key definition

REVISED ☐

Term	Definition
Specific heat capacity	The amount of heat needed to raise the temperature of a body by 1 °C.

Equatorial climate PAGES 156–157

REVISED ☐

Equatorial areas have a very distinct climate:

- annual temperatures are high (26–27 °C), being close to the Equator
- seasonal ranges are low (1–2 °C), but diurnal (daytime) ranges are greater (10–15 °C)
- rainfall is high (more than 2 000 mm per year), convectional in nature and occurs throughout the year.

Hot desert climates

REVISED ☐

Desert climates are very severe. Typical characteristics include:

- high daytime temperatures (30–40 °C) throughout the year
- a large temperature difference, often as much as 50 °C, between day and night
- low, unreliable and irregular rainfall (about 250 mm per year).

Factors affecting climate PAGES 157–159

Latitude

Areas that are close to the Equator receive more heat than areas that are close to the poles. This is due to two reasons:

1 Incoming solar radiation (insolation) is concentrated near the Equator, but dispersed near the poles.

2 Insolation near the poles has to pass through a greater amount of atmosphere and there is more chance of it being reflected back out to space.

Altitude

Temperature decreases with altitude. On average it drops about 1 °C for every 100 m. That means 10 °C over 1000 m. This is because air at higher altitudes is thinner and less dense.

Pressure systems

Low-pressure systems are associated with uplift (rising) of air, condensation, cloud formation and rain – as in a rainforest. In contrast, high-pressure systems are associated with descending (sinking) air, and dry conditions, such as in hot deserts.

Distance from the sea

It takes more energy to heat up water than it does to heat land. This is known as the **specific heat capacity** (you may come across this term in science, too). However, it takes longer for water to lose heat. Hence, land is hotter than the sea by day, but colder than the sea by night. Places that are close to the sea are cool by day, but mild by night. With increasing distance from the sea this effect is reduced.

Prevailing winds

Prevailing winds are the most frequent winds in an area. Their effect depends upon where they come from. The south-west winds that affect the British Isles bring warm air from the mid Atlantic. By contrast, north-east winds from Siberia bring bitterly cold conditions in winter.

Ocean currents

The effect of an ocean current depends upon whether it is a cold ocean current or a warm ocean current. Cold ocean currents lower the temperature of the coastlines they affect. This is very noticeable with the Labrador current, which reduces the temperature of coastal areas in north-east North America. Cold currents in tropical areas, for example, the Benguela current off the coast of Namibia, can cause deserts to form because they produce very little rain.

Aspect

Aspect is the direction a place faces. On a local scale aspect is very important. In the northern hemisphere south-facing places are warmer than north- and east-facing places.

Test yourself

1 Compare the main climate characteristics of:
  a equatorial climates
  b hot desert climates
2 Briefly explain how climate is affected by:
  a latitude
  b altitude

Answers on page 126

Tropical rainforests PAGES 159–160

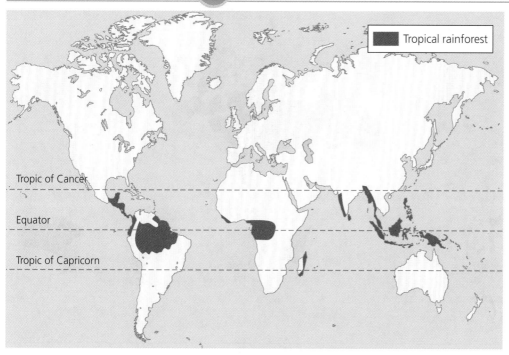

▲ **Figure 2.21** World distribution of tropical rainforests

Vegetation in the rainforest is well adapted to the climatic conditions (see the textbook, Figure 2.99 page 160). Photosynthesis and growing can take place all year. The vegetation is layered, and the shape of the crowns varies with the layer, in order to receive light. There are up to 200 species of tree per hectare (an area the size of a rugby pitch) including fig, teak, mahogany and rosewood.

Over 50 per cent of all animal species on Earth are found in tropical rainforest areas (6 per cent of the land), and most of these are arboreal (live in trees). Many species are still to be identified – between 2010 and 2013 over 440 new species were identified in the Amazon alone.

However, tropical soils are very deep and infertile. The heavy rain washes out much of the clay, salts and other nutrients. Decaying vegetation is broken down rapidly under hot, wet conditions and releases some nutrients back into the soil. As a result of the poor soil, plants that grow throughout the year take up these nutrients before they are washed away. This is called nutrient recycling. The cycle is broken when the rainforest is disturbed. This could be through fires or cutting trees down.

Tip

When writing about ecosystems, give specific details (for example, mean temperature, rainfall total, names of selected plants and animals), rather than a generalised account that could refer to any ecosystem.

It is illegal to photocopy this page

Some areas never recover from this, and the rainforest disappears gradually. This means that tropical rainforests cannot support many people, and that any part of the rainforest can become infertile quickly. However, more and more people are living in the rainforest, so the pressures are increasing. The area has a year-round growing season, but it is limited in the number of people it can support.

Hot deserts PAGES 164–167

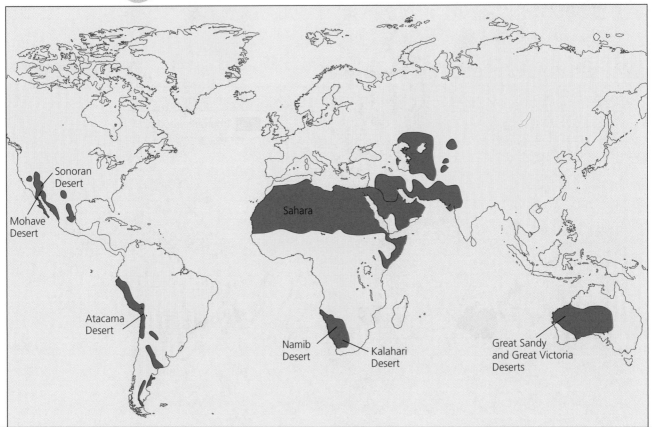

▲ **Figure 2.22** World distribution of hot deserts

Hot deserts are found in a number of locations (Figure 2.22):

- in subtropical areas between 20° and 30° north and south of the Equator (for example, the Sahara)
- in areas of rain shadow (for example, Patagonia)
- at great distance from the sea (for example, parts of the Central Australian desert)
- where there are cold, upwelling offshore sea currents (for example, the Namib desert in Namibia).

Soils are poor because they contain very few nutrients and organic matter and because they are very dry.

Vegetation from desert margins is often referred to as scrub. Plants adapt to life in hot deserts by:

- growing long water-seeking roots
- producing only a few leaves, to reduce transpiration (moisture loss)

- storing water (the cactus is a good example)
- producing seeds that wait for rainfall before growing and completing a very quick life cycle.

Typical species include saguaro cactus, prickly pear, creosote bush and palo verde.

Animals are adapted to living in the desert by:

- being nocturnal
- remaining underground for much of the day
- having large ears/panting to reduce body heat
- becoming dormant during periods of extreme heat and moisture shortages.

Test yourself

3 Briefly describe the vegetation characteristics of tropical rainforests.

4 Explain how plants are adapted to hot desert environments.

Answers on page 126

Causes and effects of tropical rainforest deforestation 📖 PAGES 161–163

REVISED

There are a large number of effects of deforestation including:

- disruption to the circulation and storage of nutrients
- surface erosion and compaction of soils
- sandification
- increased flood levels and sediment content of rivers
- climatic change
- loss of biodiversity.

Deforestation disrupts the closed system of nutrient cycling within tropical rainforests. Inorganic elements are released through burning and are quickly flushed out of the system by the high-intensity rains.

Soil erosion is also associated with deforestation. As a result of soil compaction, there is a decrease in infiltration and an increase in overland runoff and surface erosion.

Sandification is a process of selective erosion. Raindrop impact washes away the finer particles of clay and humus, leaving behind the coarser and heavier sand.

As a result of the intense surface runoff and soil erosion, rivers have a higher flood peak and a shorter time lag. However, in the dry season river levels are lower, the rivers have greater turbidity (murkiness due to more sediment), an increased bed load, and carry more silt and clay in suspension.

Other changes relate to climate. As deforestation progresses, there is a reduction of water that is re-evaporated from the vegetation, hence the recycling of water must diminish.

Case study: Danum Valley Conservation Area

Danum Valley Conservation Area (DVCA) contains more than 120 mammal species, including 10 species of primate. DVCA and the surrounding forest form an important reservation for orangutans, as well as being particularly rich in other large mammals, including the Asian elephant, Malayan sun bear, clouded leopard, bearded pig and several species of deer. Over 340 species of bird have been recorded at Danum, including the argus pheasant, Bulwer's pheasant, and seven species of pitta bird. The area also provides one of the last refuges for the endangered Sumatran rhinoceros.

DVCA covers 43 800 hectares, comprising almost entirely lowland dipterocarp forest (dipterocarps are valuable hardwood trees). It is the largest expanse of pristine forest of this type remaining in Sabah, north-east Borneo.

Until the late 1980s, the area was under threat from commercial logging. The establishment of a long-term research programme between Yayasan Sabah and the Royal Society in the UK created local awareness of the conservation value of the area and provided important scientific information about the forest and what happens to it when it is disturbed through logging. Danum Valley is controlled by a management committee representing all the relevant local institutions, including wildlife, forestry and commercial sectors. To the east of DVCA is the 30 000-hectare Innoprise-FACE Foundation Rainforest Rehabilitation Project (INFAPRO), one of the largest forest rehabilitation projects in South East Asia, which is replanting areas of heavily disturbed logged forest.

Because all areas of conservation and replanting are embedded within the larger commercial forest, the value of the whole area is greatly enhanced. Movement of animals between forest areas is enabled and allows the continued survival of some important and endangered Borneo animals such as the Sumatran rhinoceros, the orang-utan and the Borneo elephant.

In the late 1990s, a hotel opened on the north-eastern edge of DVCA. It has established flourishing ecotourism in the area and exposed this unique forest to a wider range of visitors than was previously possible. As well as raising revenue for the local area, it has raised the international profile of the area as an important centre for conservation and research.

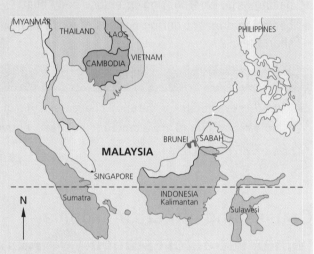

▲ **Figure 2.23** Map of Malaysia

▲ **Figure 2.24** Map of Sabah

Case study: The Sonoran Desert

The Sonoran Desert is located in southern USA (southern California and southern Arizona) and northern Mexico.

Its vegetation includes the saguaro cactus, which can grow to a height of 15 m and live for up to 175 years. Its ribbed stem expands as it fills with water during the winter wet season. Its stem shape reduces wind speed and water loss from the plant, while sunken stomata also reduce water loss. It has shallow roots to catch water from storms before it evaporates. The palo verde is a small, drought-tolerant tree that loses its leaves in the dry season, but its green bark allows it to photosynthesise without leaves. Creosote bushes have small, dark leaves to reduce transpiration. Plant density depends upon water availability.

Soils in the Sonoran Desert are typically thin, relatively infertile and alkaline. Seasonal rains carry soluble salts down through the soil. However, during the dry season these are drawn up to surface by evaporation. Concentrations may become toxic for some plants. In addition, flash flooding can compact the soil, leaving the surface impermeable.

There has been considerable human impact in the area. Some cities, notably Phoenix in Arizona, have expanded rapidly at the expense of the desert. The increased demand for, and abstraction of, water has lowered water tables. Road construction and pipelines have affected the movement of mammals, and fenced highways have prevented pronghorn antelopes, for example, from reaching water supplies. Domesticated animals have escaped into the wild and reduced grazing availability for wild mammals. The introduction of exotic plant species, such as tamarisk, has displaced native species such as cottonwoods and desert willows.

Exam-style questions

1 Briefly explain why tropical rainforests are hot and wet. [4]
2 Explain why soils in the tropical rainforest are usually infertile. [3]
3 Describe the main difficulties in developing hot, arid areas. [3]

Answers on page 131

3.1 Development

Key objectives

You should be able to:

- use a variety of indicators to assess the level of development of a country
- identify and explain inequalities between and within countries
- classify production into different sectors and give illustrations of each
- describe and explain how the proportions employed in each sector vary according to the level of development
- describe and explain the process of globalisation, and consider its impacts.

Case study required

- A transnational corporation and its global links.

Key definitions

REVISED

Term	Definition
Development	The use of resources to improve the quality of life in a country.
Gross National Product (GNP)	The total value of goods and services produced by a country in a year, plus income earned by the country's residents from foreign investments and minus income earned within the domestic economy by overseas residents.
Gross National Product per capita	The total GNP of a country divided by the total population.
Development gap	The differences in wealth, and other indicators, between the world's richest and poorest countries.
Human Development Index (HDI)	Combines four indicators of development: life expectancy at birth; mean years of schooling for adults aged 25 years; expected years of schooling for children of school entering age; GNI per capita (PPP$).
Least developed countries (LDCs)	The poorest of the developing countries. They have major economic, institutional and human resource problems.
Newly industrialised countries (NICs)	Nations that have undergone rapid and successful industrialisation since the 1960s.
Gini coefficient	Technique used to show the extent of income inequality.
Cumulative causation	The process whereby a significant increase in economic growth can lead to even more growth as more money circulates in the economy.
Formal sector	That part of an economy known to the government department responsible for taxation and to other government offices.
Informal sector	That part of the economy operating outside official recognition.
Product chain	The full sequence of activities needed to turn raw materials into a finished product.
Globalisation	The increasing interconnectedness and interdependence of the world economically, culturally and politically.
Transnational corporation (TNC)	A firm that owns or controls productive operations in more than one country through foreign direct investment (FDI).
Diffusion	The spread of a phenomenon over time and space.
Internet	A group of protocols by which computers communicate.
New international division of labour (NIDL)	Divides production into different skills and tasks that are often spread across a number of countries.

Indicators of development PAGES 172–176

Development, or improvement in the quality of life, includes increasing wealth, but also involves other important aspects of our lives (Figure 3.1). For example, development occurs in a low-income country when:

- local food supply improves due to investment in farm machinery and fertilisers
- the electricity grid extends outwards from the main urban areas to rural areas.

Gross National Product

One of the traditional indicators of a country's wealth is the **Gross National Product** (GNP). To take account of the different populations of countries the **Gross National Product per capita** is often used. Figure 3.3 on page 173 of the textbook shows GNP per capita of countries for 2013. It is clear to see where regions of high and low GNP per capita are located. The **development gap** between the world's wealthiest and poorest countries is huge.

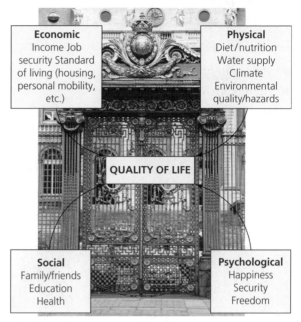

Economic Income Job security Standard of living (housing, personal mobility, etc.)

Physical Diet/nutrition Water supply Climate Environmental quality/hazards

QUALITY OF LIFE

Social Family/friends Education Health

Psychological Happiness Security Freedom

▲ **Figure 3.1** Factors comprising the quality of life

▼ **Table 3.1** Important indicators of development

Literacy	Adult literacy is an important measure of educational standards. In 2015, the global adult literacy rate was 85 per cent, but in over a dozen African countries, literacy rates are below 50 per cent. Female literacy is particularly important because so many aspects of development depend on it.
Life expectancy	Life expectancy is, to a large extent, the end result of all the factors contributing to the quality of life in a country. Rates of life expectancy have converged significantly between rich and poor countries during the last 50 years, in spite of a widening wealth gap.
Infant mortality	The infant mortality rate is an important measure of health inequality. Fortunately, infant mortality rates have fallen sharply in many developing countries over the last 20 years. However, the infant mortality rate in Africa is more than 11 times that of Europe.

Other measures of development include:

- doctors per 100 000 people
- energy consumption per capita
- percentage of the population living in urban areas
- internet penetration rate.

The Human Development Index: a broader measure of development

The **Human Development Index** (HDI) was devised by the United Nations in 1990. The current index combines four indicators of development:

- life expectancy at birth
- mean years of schooling for adults aged 25 years
- expected years of schooling for children of school entering age
- GNI per capita (PPP$, namely purchasing power parity, i.e. how much you can buy for your income related to local prices).

Tip

It is important to understand the difference between economic growth and development. The former is an increase in GDP while development is a more wide-ranging concept concerning many more aspects of the quality of life.

No single measure can provide a complete picture of the differences in development between countries. This is why the United Nations combines four measures for the HDI. Although the development gap can be measured in a variety of ways it is generally taken to be increasing.

The HDI divides the countries of the world into four groups (Figure 3.2):

- Very high human development
- High human development
- Medium human development
- Low human development.

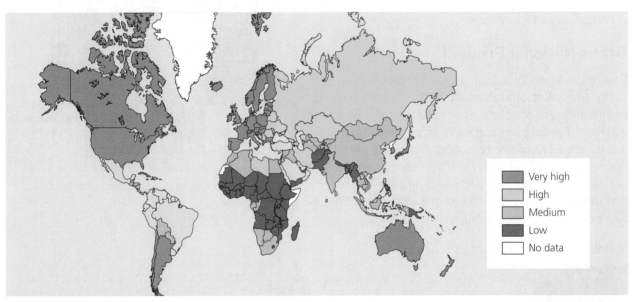

▲ **Figure 3.2** Map of the Human Development Index, 2011

Explaining inequalities between countries

 **PAGES 176–178**

REVISED

Stages of development

A reasonable division of the world in terms of economic development is shown in Figure 3.3.

The **least developed countries** (LDCs) are the poorest of the developing countries. Their problems are often made worse by geographical handicaps such as very low rainfall and natural and man-made disasters. At present 48 countries are identified as LDCs. Of these 34 are in Africa.

Newly industrialised countries (NICs) are nations that have moved up the development ladder, having previously been considered developing countries. The first countries to become newly industrialised countries (in the 1960s) were South Korea, Singapore, Taiwan and Hong Kong. The media referred to them as the 'four Asian tigers'. A 'tiger economy' is one that grows very rapidly.

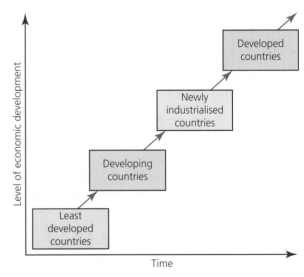

▲ **Figure 3.3** Stages of development

The success of these four countries provided a model for others to follow, such as Malaysia, Brazil, China and India. In the last 20 years the growth of China has been particularly impressive. South Korea and Singapore have developed so much that many people now consider them to be developed countries.

Explaining the development gap

There has been much debate about the causes of development. Reasons for variations between countries include:

- **Physical geography**: for example, landlocked countries have generally developed more slowly than coastal nations, and tropical countries have grown more slowly than those in temperate latitudes.
- **Economic policies**: for example, open economies which welcomed and encouraged foreign investment have developed faster than closed economies; institutional quality in terms of good government, law and order and lack of corruption generally result in high rates of growth.
- **Demography**: progress through demographic transition is a major factor, with the highest rates of economic growth experienced by those nations where the birth rate had fallen the most.

Consequences of the development gap

The development gap has significant consequences for people in the most disadvantaged countries, as shown in Table 3.2.

▼ **Table 3.2** The consequences of poverty

Economic	In 2015, about 10 per cent of the world's population lived on less than $2 per day. Poor countries frequently lack the ability to pay for food, agricultural innovation and investment in rural development.
Social	About 780 million people in poor countries cannot read or write. Over 660 million people do not have access to clean water and 2.4 billion to basic sanitation.
Environmental	Poor countries experience increased vulnerability to natural disasters. They lack the capacity to adapt to droughts induced by climate change. Poor farming practices lead to environmental degradation.
Political	Poor countries that are low on the development scale often have non-democratic governments or they are democracies that function poorly.

Explaining inequalities within countries PAGES 179–183 REVISED

- The **Gini coefficient** is a technique frequently used to show the extent of income inequality.
- It is defined as a ratio with values between 0 and 1.0.
- A low value indicates a more equal income distribution while a high value shows more unequal income distribution.
- In general, more affluent countries have a lower income gap than lower income countries.
- Southern Africa and South America show up clearly as regions of very high income inequality. Europe is the world region with the lowest income inequality.

A theory of regional disparities

The model of **cumulative causation** helps to explain regional disparities. Figure 3.4 is a simplified version of the model. There are three stages of regional disparity:

- the pre-industrial stage when regional differences are small
- a period of rapid economic growth with increasing regional economic divergence
- a stage of regional economic convergence.

In the model, economic growth begins with the location of new manufacturing industry in the region with the best combination of advantages. Once growth begins in this 'core' region, flows of labour, capital and raw materials develop to support it. The growth region undergoes further expansion by the cumulative causation process. A detrimental negative effect (the backwash effect) is transmitted to the less developed regions (the periphery) as skilled labour and locally generated capital are attracted away. Manufactured goods and services produced under the economies of scale of the core region undercut smaller-scale enterprises in the periphery. The wealth gap between the core and the periphery widens and regional inequality increases (regional economic divergence).

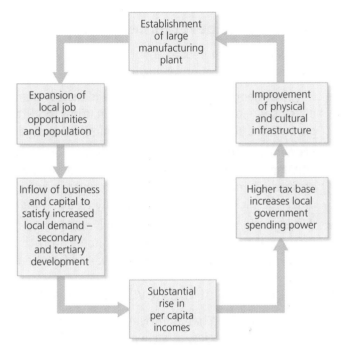

▲ **Figure 3.4** Simplified model of cumulative causation

However, increasing demand for raw materials from resource-rich parts of the periphery may stimulate growth in such regions. This may set off the process of cumulative causation, leading to the development of new centres of self-sustained economic growth (spread effects). If the process is strong enough, the inequality between core and periphery may begin to narrow.

Many developing countries are in the stage where the wealth gap between core and periphery is still widening. Thus, they have a high Gini coefficient.

Factors affecting inequalities within countries

- **Residence**: where people live can have a big impact on their quality of life. This includes: regional differences within countries; urban/rural disparities; and intra-urban contrasts.
- **Ethnicity**: some ethnic groups in a population often have income levels far below the dominant group(s) in the same population. This is often the result of discrimination.
- **Employment**: jobs in the **formal sector** of the economy generally provide better pay and greater security than jobs in the **informal sector**.
- **Education**: higher levels of education generally lead to better-paid employment. In developing countries there is a clear link between education levels and family size, with those with the least education having the largest families.
- **Land ownership**: the distribution of land ownership (tenure) has had a major impact on disparities in many countries. The greatest disparities tend to occur alongside the largest inequities in land ownership.

Test yourself

1 List the indicators used in the human development index (HDI).

2 How many countries are identified as LDCs?

3 What is the *Gini coefficient*?

Answers on page 126

Classifying production into different economic sectors 📖 PAGES 183-184

- The **primary sector** exploits raw materials from land, water and air. Farming, fishing, forestry, mining and quarrying make up most of the jobs in this sector.
- The **secondary sector** manufactures primary materials into finished products. Activities in this sector include the production of processed food, furniture and motor vehicles.
- The **tertiary sector** provides services to businesses and to people. Retail employees, drivers, teachers and nurses are examples of occupations in this sector.
- The **quaternary sector** uses high technology to provide information and expertise. Research and development is an important part of this sector. Jobs in this sector include aerospace engineers, research scientists and biotechnology workers.

> **Tip**
>
> You should take care with the word 'industry' as it can be applied to all sectors of the economy (for example, the agricultural industry and the service industry).

The **product chain** can be used to illustrate the four sectors of employment. The food industry provides a good example (Figure 3.5).

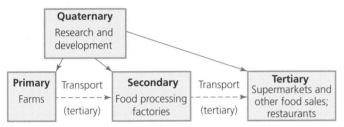

▲ **Figure 3.5** The food industry's product chain

How employment structure varies 📖 PAGES 184-185

As an economy develops, the proportion of people employed in each sector changes (Figure 3.6). Countries such as the USA and the UK are 'post-industrial societies' where most people are employed in the tertiary sector. Yet in 1900, 40 per cent of employment in the USA was in the primary sector. However, mechanisation drastically reduced the demand for labour in primary industries. As these jobs disappeared, people moved to urban areas where secondary and tertiary employment was expanding. Only about 2 per cent of employment in the USA is now in the primary sector.

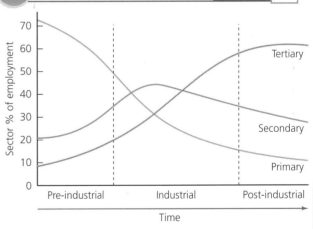

▲ **Figure 3.6** The sector model

Human labour has been steadily replaced in manufacturing too. In more and more factories, robots and other advanced machinery handle assembly-line jobs that once employed large numbers of people. The tertiary sector is also changing as computer networks have reduced the number of people required in some occupations. But elsewhere, service employment is often rising, such as in health and tourism. In developed countries employment in the quaternary sector has become more and more important. Employment in the quaternary sector is a significant measure of how advanced an economy is.

Table 3.3 compares the employment structure of a developed country, a newly industrialised country and a developing country.

There is a very clear link between employment structure and indicators of development. A graphical method often used to compare the employment structure of a large number of countries is the triangular graph.

▼ **Table 3.3** Employment structure of a developed country, an NIC and a developing country

Country	% primary	% secondary	% tertiary
Australia (developed)	4	21	75
Malaysia (NIC)	11	36	53
Bangladesh (developing)	45	30	25

The process of globalisation PAGES 186–188

REVISED

Globalisation is a term that describes rapidly increasing global links. Most political borders are not the obstacles they once were and as a result goods, capital, labour and ideas flow more freely across them than ever before.

Transnational corporations

- TNCs can exploit raw materials, produce goods such as cars and oil, and provide services such as banking. The 100 largest TNCs represent a significant proportion of total global production.

- TNCs and nation states (countries) are the two main elements of the global economy. The governments of countries individually and collectively set the rules for the global economy, but the bulk of investment is through TNCs.

- Under this process, manufacturing industry at first, and more recently services, have relocated in significant numbers from developed countries to selected developing countries as TNCs have taken advantage of lower labour costs and other ways to reduce costs. It is this process which has resulted in the emergence of an increasing number of newly industrialised countries since the 1960s.

- Twenty years ago the vast majority of the world's TNCs had their headquarters in North America, Western Europe and Japan. However, over the last two decades NICs such as South Korea, China and India have been accounting for an increasing slice of the global economy. Much of this economic growth has been achieved through the expansion of their own most important companies.

- TNCs have a huge impact on the global economy in general and in the countries in which they choose to locate in particular. They play a major role in world trade in terms of what and where they buy and sell.

The role of technology

Advances in technology have affected all aspects of global economic activity. Major advances in transportation and telecommunications systems have significantly reduced the geographical barriers separating countries and peoples. As time has progressed, the **diffusion** of new ideas has speeded up.

The **internet** has been essential to the development and speed of globalisation. It is the fastest-growing mode of communication ever. It has been estimated that the number of internet users around the world

Test yourself

4 Define the *product chain*.

5 What is a *TNC*?

6 State the increase in the number of global internet users between 2000 and 2016.

Answers on page 126

increased from 361 million in 2000 to 3.4 billion in 2016. The internet has allowed TNCs to manage complex operations all over the world. TNCs can react more quickly than ever before to changing consumer demand.

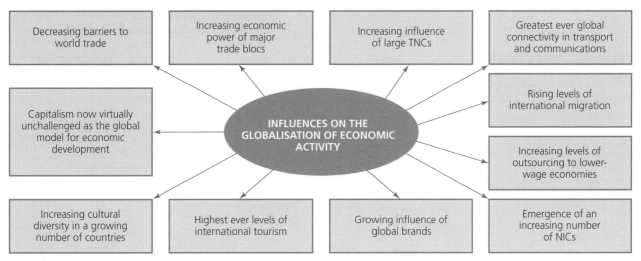

▲ **Figure 3.7** Influences on the globalisation of economic activity

Other factors responsible for economic globalisation

Figure 3.7 shows the main influences on the globalisation of economic activity. Until the post-1950 period, industrial production was mainly organised within individual countries. This has changed rapidly in the last 60 years or so with the emergence of a **new international division of labour** (NIDL). The NIDL divides production into different skills and tasks that are often spread across a number of countries.

The impacts of globalisation PAGES 189–193 REVISED

Table 3.4 shows some of the impacts of globalisation at global, national and local scales. Table 3.5 summarises some of the costs and benefits of globalisation to the UK.

▼ **Table 3.4** Examples of the impacts of globalisation at global, national and local scales

Global	National	Local
The growing power of TNCs and global brands	Concerns about loss of sovereignty to regional and international organisations	Small local businesses often find it difficult to compete with major global companies
The emergence of an increasing number of NICs	Increased cultural diversity from international migration	Closure of a TNC branch plant can cause high local unemployment
Development of a hierarchy of global cities	Higher levels of incoming and outgoing international tourism	The populations of many local communities have become more multicultural
The emergence of powerful trade blocs	TNCs employing an increasing share of the workforce	Families now more likely to be spread over different countries due to increased international migration
Environmental degradation caused by increasing economic activity	Increasing incidences of trans-boundary pollution	The development of 'ethnic villages' in large urban areas

▼ **Table 3.5** The costs and benefits of globalisation to the UK

Perspective	Benefits	Costs
Economic	As one of the world's most 'open' economies, the UK attracts a very high level of foreign direct investment, creating employment and contributing to GDP; a high level of investment abroad by UK companies also increases national income. Financial deregulation has enhanced the position of the City as one of the world's top three financial centres. Low-cost manufactured goods from China and elsewhere have helped keep inflation low.	High job losses in traditional industries due to global shift and deindustrialisation. TNCs can move investment away from the country as quickly as they can bring it in, causing loss of jobs and corporation tax. Speculative investment, causing economic uncertainty, has increased with financial deregulation. There is a widening gap between the highest- and lowest-paid workers.
Social	Economic growth has allowed high levels of spending on education and health in particular. Globalisation is a large factor in the cosmopolitan nature of UK society. The transport and communications revolution has transformed lifestyles.	A strong economy has attracted a very high level of immigration in recent years, with increasing concerns that this is unsustainable.
Political	Strong trading relationships with a large number of other countries brings political influence.	Voter apathy as many people see loss of political power to the EU and major TNCs. International terrorism is a growing threat with increasing ethnic diversity, rapid transportation and more open borders.
Environmental	Deindustrialisation has improved environmental conditions in many areas; increasing international cooperation to solve cross-border environmental issues gives a better chance of such problems being addressed.	Population growth has an impact on the environment, with increasing demand for land, water and other resources. Rapid industrial growth in China and elsewhere has an impact on the global environment, including the UK.

Case study: Tata Group and its global links

- Indian companies are becoming increasingly transnational in their operations.
- Tata is India's best-known global brand. With its headquarters in Mumbai, it encompasses seven business sectors.
- Tata Group has over 100 companies with each of them operating independently. Some of the largest of these companies are Tata Steel, Tata Motors, Tata Chemicals and Tata Global Beverages.
- In recent decades Tata has expanded rapidly around the world. Tata Group now has operations in more than 100 countries and receives more than 60 per cent of its revenue from outside India. In 2016, the total number of employees worldwide was 666 000.
- Tata Group has steadily moved up the 'value chain' by producing more sophisticated and higher-value products.
- Tata has a considerable presence in the UK. Key acquisitions have included: Corus Group by Tata Steel for $13 billion in 2007; Jaguar and Land Rover by Tata Motors for $2.5 billion in 2008.
- The objective has often been to buy world-renowned brands that are synonymous with high quality.
- However, some Tata companies in the UK have encountered financial problems in recent years.
- Tata Group has set great store by its reputation for social responsibility. Tata was awarded the Carnegie Medal for Philanthropy in 2007.

Sample exam question

a Which activities make up the primary sector of an economy? [2]

Student's answer

> *a* The primary sector exploits the raw materials in a country. The main economic activities in the primary sector are farming, fishing, forestry, mining and quarrying.

Teacher's comments

A clear and concise answer, gaining the two marks available. The answer comprises a good opening statement followed by relevant and accurate elaboration.

b Why does the primary sector dominate employment in the poorest countries of the world? [3]

Student's answer

> *b* The poorest countries of the world have more than 70% of their employment in the primary sector. Lack of investment in general means that agriculture and other areas of the primary sector are very labour intensive and jobs in the secondary and tertiary sectors are limited in number.

Teacher's comments

The opening sentence includes a useful statistic showing the extent of the dominance of the primary sector in poor countries. The following sentence provides the necessary explanation in terms of both the primary sector and the other sectors of the economy. The student gains the full three marks available.

c Explain the changes in employment structure that have occurred in NICs. [4]

Student's answer

> *c* In NICs such as China and Brazil employment in manufacturing has risen rapidly in recent decades. NICs have attracted high levels of foreign direct investment from transnational corporations. This has not just been in manufacturing, but in the service sector in some countries such as India. The increasing wealth of NICs allows for greater investment in agriculture. This includes mechanisation, which results in falling demand for labour on the land. So, as employment in the secondary and tertiary sectors rises, employment in the primary sector falls.

Teacher's comments

This answer shows clear knowledge and understanding of employment changes in the different sectors in NICs. Relevant use of examples adds to the quality of the answer. The student gains all four marks available.

Exam-style questions

1 **a** Define the primary sector of an economy. [2]
   b Why does the primary sector dominate employment in the poorest countries of the world? [3]
   c Explain the changes in employment structure that have occurred in NICs. [4]
2 **a** Define a transnational corporation. [2]
   b Describe and explain the role of transnational corporations in the global economy. [6]

Answers on pages 131–132

3.2 Food production

Key objectives

You should be able to:

- describe and explain the main features of an agricultural system: inputs, processes and outputs
- recognise the causes and effects of food shortages and describe possible solutions to this problem.

Case studies required

- A farm or agricultural system.
- A country or region suffering from food shortages.

Key definitions

REVISED

Term	Definition
System	A practice in which there are recognisable inputs, processes and outputs.
Irrigation	Supplying dry land with water by systems of ditches and also by more advanced means.
Economies of scale	The reduction in unit cost as the scale of an operation increases.
Agricultural technology	The application of techniques to control the growth and harvesting of animal and vegetable products.
Land tenure	The ways in which land is or can be owned.
Green Revolution	The introduction of high-yielding seeds and modern agricultural techniques in developing countries.

Agricultural systems PAGES 194–196

REVISED

Individual farms and general types of farming can be seen to operate as a **system** (Figure 3.8). A farm requires a range of inputs such as labour so that the processes that take place on the farm, such as harvesting, can be carried out. The aim is to produce the best possible outputs such as milk, eggs and crops. A profit will only be made if the income from selling the outputs is greater than the costs of the inputs and processes.

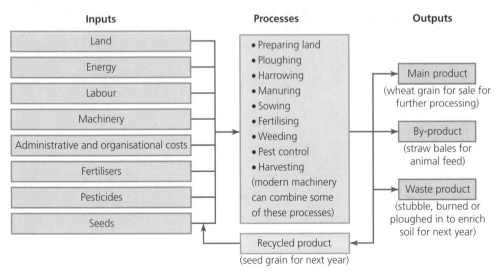

▲ **Figure 3.8** Systems diagram for a wheat farm

Different types of agricultural system can be found within individual countries and around the world. The most basic distinctions are between:

- arable, pastoral and mixed farming
- extensive and intensive farming
- subsistence and commercial farming
- organic and non-organic farming.

▼ **Table 3.6** Farming types and their characteristics

Farming type	Characteristics
Arable farming	Cultivation of crops and not involved with livestock. An arable farm may concentrate on one crop (monoculture), such as wheat, or may grow a range of different crops.
Pastoral farming	Involves keeping livestock such as dairy cattle, beef cattle, sheep and pigs.
Mixed farming	Involves cultivating crops and keeping livestock together on a farm. Usually on a mixed farm at least part of the crop production will be used to feed the livestock.
Subsistence farming	The most basic form of agriculture where the produce is consumed entirely or mainly by the family who work the land or tend the livestock.
Commercial farming	The objective is to sell everything that the farm produces. The aim is to maximise yields in order to achieve the highest profits possible.
Extensive farming	A relatively small amount of agricultural produce is obtained per hectare of land, so such farms tend to cover large areas of land. Inputs per unit of land are low.
Intensive farming	Characterised by high inputs per unit of land to achieve high yields per hectare. Examples of intensive farming include market gardening, dairy farming and horticulture.
Organic farming	Does not use manufactured chemicals, so production is without chemical fertilisers, pesticides and herbicides. Instead, animal and green manures are used along with mineral fertilisers such as fish and bone meal.

The influence of natural and human inputs on agricultural land use 📖 PAGES 196–200   REVISED

North America has many different physical environments, allowing a wide variety of crops to be grown and livestock kept. New technology and high levels of investment have steadily extended farming into more difficult environments. However, there are certain things that technology and investment can do little to alter. So relief, climate and soils set broad limits as to what can be produced. This leaves the farmer with some choices, even in difficult environments. The farmer's decisions are then influenced by economic, social/cultural and political factors.

Physical factors

- Temperature is a critical factor as each type of crop requires a minimum growing temperature and a minimum growing season.
- Latitude, altitude and distance from the sea are the major influences on temperature.
- Precipitation is equally important: not just the annual total but the way it is distributed throughout the year.
- Soil type and fertility have a huge impact on agricultural productivity.
- Locally, aspect and the angle of slope may also be important factors in deciding how to use the land.
- Water is vital for agriculture. **Irrigation** is an important factor in farming in many parts of the world.

Economic factors

- The cost of growing different crops or keeping different livestock varies. The market prices for agricultural products will vary also and can change from year to year.

- The necessary investment in buildings and machinery can mean that some changes in farming activities are very expensive. Thus, it is not always easy for farmers to react quickly to changes in consumer demand.
- In most countries there has been a trend towards fewer but larger farms. Large farms allow **economies of scale** to operate, which reduce the unit costs of production.
- The cost of transporting farm produce to market can be a big influence on what is produced on a farm.
- The status of a country's **agricultural technology** is vital for its food security and other aspects of its quality of life.

Social/cultural factors

- What a farm has produced in the past can be a significant influence on current farming practices. Tradition matters more in some farming regions than others.
- **Land tenure** means the ways in which land is or can be owned. In the past inheritance laws have had a huge impact on the average size of farms.
- In most societies women have very unequal access to, and control over, rural land and associated resources.

Political factors

The influence of government on farming has steadily increased in many countries. For example, in the USA the main parts of government farm policy over the past half century have been price support loans, production controls and income supplements. An agricultural policy can cover more than one country, as evidenced by the EU's Common Agricultural Policy.

Test yourself

1 List the **three** sub-sections of a system.
2 Define *pastoral farming*.
3 What is *land tenure*?

Answers on page 126

Case study: An agricultural system – intensive rice production in the Lower Ganges Valley

▼ Table 3.7

Physical conditions	Suitable for rice cultivation because: • temperatures of 21 °C and over throughout the year allow two crops to be grown annually; rice needs a growing season of only 100 days • monsoon rainfall over 2000 mm provides sufficient water for the fields to flood, which is necessary for wet rice cultivation • rich alluvial soils are built up through regular flooding • an important dry period allows for harvesting the rice.
A water-intensive staple crop	Production systems are extremely water intensive. Much of Asia's rice production is intensive subsistence cultivation where the crop is grown on very small plots of land using a very high input of labour. 'Wet' rice is grown in the fertile silt and flooded areas of the lowlands while 'dry' rice is cultivated on terraces on the hillsides.
The farming system	Paddy fields characterise lowland production. At first, rice is grown in nurseries, then transplanted when the monsoon rains flood the paddy fields. The main rice crop is harvested when the drier season begins in late October. A second rice crop can then be planted in November. Water buffalo are used for work. The labour-intensive nature of rice cultivation provides work for large numbers of people. A high labour input is needed to: • build the embankments that surround the fields • construct irrigation canals • plant nursery rice, plough the paddy field, transplant the rice from the nursery to the paddy field, weed and harvest the mature rice crop • cultivate other crops in the dry season and possibly tend a few livestock. Rice seeds are stored from one year to provide the next year's crop.

Causes and effects of food shortages **PAGES 200–202** REVISED

▼ **Table 3.8** Causes of food shortages

Natural factors	Economic and political factors
Soil exhaustion	Low capital investment
Drought	Rapidly rising population
Floods	Poor distribution/transport difficulties
Tropical cyclones	Conflict situations
Pests	Grain reserves at very low levels
Disease	Rising food prices

In late 2012, the UN warned of an imminent worldwide food crisis, highlighting three major problems: (a) global grain reserves at critically low levels; (b) rising food prices creating unrest in many countries; (c) extreme weather resulting in climate being 'no longer reliable'. The impact of such problems has been felt most intensely in developing countries, where adequate food stocks to cover emergencies affecting food supply usually do not exist.

Short-term and long-term effects

The effects of food shortages are both short and longer term. Malnutrition can affect a considerable number of people, particularly children, within a relatively short period when food supplies are reduced. With malnutrition people are less resistant to disease and more likely to fall ill. Malnutrition reduces people's capacity to work. This is threatening to lock parts of the developing world into an endless cycle of ill health, low productivity and underdevelopment.

> **Tip**
>
> A simple, but clearly labelled sketch map can considerably enhance the presentation of a case study.

Case study: A region suffering from food shortages – Sudan and South Sudan

- The countries of Sudan and South Sudan, which were the single country of the Sudan until 2011, have suffered food shortages for decades.

- The long civil war and drought have been the main reasons for famine in the Sudan, but there are many associated factors as well (Figure 3.9, page 78).

- One of the big issues between the two sides in the civil war was the sharing of oil wealth between the government-controlled north and the south of the country where much of the oil is found.

- The United Nations has estimated that up to 2 million people were displaced by the civil war and more than 70 000 people died from hunger and associated diseases.

- At times, the UN World Food Programme has stopped deliveries of vital food supplies because the situation has been considered too dangerous for the drivers and aid workers.

The separation of Sudan into two countries has not occurred easily. There has been intermittent fighting in border regions. This has undermined agricultural production. In March 2013 the World Food Programme warned that more than 4.1 million people were likely to be short of food in South Sudan in that year. In May 2016, the UN warned that more than 5 million people in South Sudan would face severe food shortage.

Physical factors	Social factors	Agricultural factors	Economic/political factors
• Long-term decline of rainfall in southern Sudan • Increased rainfall variability • Increased use of marginal land leading to degradation • Flooding	• High population growth (3%) linked to use of marginal land (overgrazing, erosion) • High female illiteracy rates (65%) • Poor infant health • Increased threat of AIDS	• Highly variable per capita food production; long-term the trend is static • Static (cereals and pulses) or falling (roots and tubers) crop yields • Low and falling fertiliser use (compounded by falling export receipts) • Lack of a food surplus for use in crisis	• High dependency on farming (70% of labour force; 37% of GDP) • Dependency on food imports (13% of consumption 1998–2000) whilst exporting non-food goods, e.g. cotton • Limited access to markets to buy food or infrastructure to distribute it • Debt and debt repayments limit social and economic spending • High military spending

Drought in southern Sudan compounds low food intake; any remaining surpluses quickly used

Shorter-term factors leading to increased Sudanese food insecurity and famine

Both reduce food availability in Sudan and inflate food prices

Conflict in Darfur reduces food production and distribution

Situation compounded by:
• Lack of government political will
• Slow donor response
• Limited access to famine areas
• Regional food shortages

▲ **Figure 3.9** Summary of causes of famine in Sudan and South Sudan

Possible solutions to food shortages **PAGES 202–204** REVISED

Food aid

There are three types of food aid:

● relief food aid
● programme food aid
● project food aid.

The USA and the EU together provide about two-thirds of global food aid deliveries. At the international level, the main organisations are the UN World Food Programme (WFP), the UN Food and Agriculture Organization (FAO) and the Food Aid Convention.

Food aid is vital to communities in many countries, particularly in Africa but also in parts of Asia and Latin America. However, it is not without controversy:

● The selling of heavily subsidised food in African countries has undermined the ability of African farmers to produce for local markets.
● Food aid is very expensive, not least because of the high transport costs involved.

There have been recent concerns that food aid may be required for even more people in the future. Steep increases in the price of food have caused big problems in a number of countries, resulting in large-scale protests. The World Bank warned that progress on development could be destroyed by rapidly rising food costs.

Common error

Error	Why it is wrong
'Food aid should not be criticised.'	Food aid is quite rightly viewed as a good thing, but the way it is done is sometimes criticised. It is important to be aware of the 'pros' and 'cons' of this issue.

The Green Revolution

Much of the global increase in food production in the last 50 years can be attributed to the Green Revolution. Although the benefits of the Green Revolution are clear, serious criticisms have also been made:

- The necessary high inputs of fertiliser and pesticide are costly in both economic and environmental terms.
- The problems of salinisation and waterlogged soils have increased with the expansion of irrigation.
- High chemical inputs have had a considerable negative effect on biodiversity.
- Ill health has increased due to contaminated water and other forms of agricultural pollution.
- Green Revolution crops are often low in important minerals and vitamins.

UNEP's options for improving food security

The United Nations Environment Programme has argued that increasing food energy efficiency provides a critical path for significant growth in food supply without compromising environmental sustainability.

Sample exam question

a What is an agricultural system? [3]

Student's answer

a An agricultural system is a type of farming such as a wheat farm or a dairy farm. Each type of farming has inputs, processes and outputs.

b With reference to examples, distinguish between intensive and extensive farming. [4]

Student's answer

b Intensive farming is characterised by high inputs per unit of land to achieve high yields per hectare. Extensive farming is where a relatively small amount of agricultural produce is obtained per hectare of land, so such farms tend to cover large areas of land. Inputs per unit of land are low.

Teacher's comments

This is a good, clear answer as far as it goes, scoring two marks out of the maximum of three. To gain the third mark the student should have given examples of inputs (e.g. labour and energy), processes (e.g. ploughing and harvesting) and outputs (e.g. crops and milk).

Teacher's comments

The student has achieved two marks out of the maximum of four. The answer clearly describes the difference between intensive and extensive farming, but there is no reference at all to examples of these types of farming. Examples of extensive farming are sheep farming in Australia and wheat cultivation on the Canadian Prairies. Examples of intensive farming are horticulture in the Netherlands and dairy farming in Denmark.

c Discuss three physical factors that affect agricultural land use. [6]

Student's answer

c Temperature is a major factor influencing farming as each type of crop requires a minimum growing temperature and a minimum growing season. Latitude, altitude and distance from the sea are the main influences on temperature.

Precipitation is another very important factor influencing the type of farming possible in a region. It is not just the annual amount of precipitation that is important, but the way it is distributed throughout the year. Long, steady periods of rainwater to infiltrate into the soil are best, making water available for crop growth throughout the year. In contrast, short, heavy downpours can result in surface runoff, leaving less water available for crop growth and also contributing to soil erosion.

A third physical factor affecting farming is soil fertility.

Teacher's comments

The student has gained five marks out of the six available. This answer shows good understanding of the influences of temperature and precipitation on agriculture. However, when it comes to the consideration of a third physical factor, the student is only able to name soil fertility, with no attempt to elaborate and gain the final mark available.

Exam-style questions

1 a How can farming be seen to operate as a system? [2]
  b Explain the difference between (i) intensive and extensive farming and (ii) subsistence farming and commercial farming. [4]

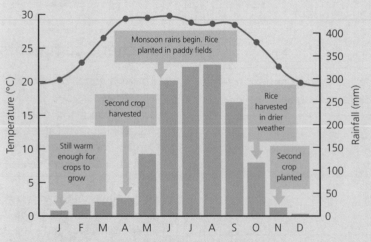

▲ **Figure 3.10** Climate graph for Kolkata, India

2 a Describe the distribution of temperature and rainfall on the graph (Figure 3.10). [3]
  b How can temperature and rainfall influence farming in a region? [5]

Answers on page 132

3.3 Industry

Key definitions

REVISED

Term	Definition
By-product	Something that is left over from the main production process, which has some value and therefore can be sold.
Waste product	All manufacturing industries produce waste product that has no value and must be disposed of. Costs will be incurred in the disposal of waste product.
Footloose industries	Industries that are not tied to certain areas because of energy requirements or other factors.
Industrial agglomeration	The clustering together of economic activities in close proximity to one another.
Industrial estate	An area zoned and planned for the purpose of industrial development.
Greenfield locations	An area of agricultural land or some other undeveloped site earmarked for commercial development or industrial projects.

Industrial systems and types PAGES 205–206

REVISED

Manufacturing industry as a whole or an individual factory can be regarded as a system. Industrial systems, like agricultural systems, have inputs, processes and outputs (Figure 3.11).

- **Inputs** are the elements that are required for the processes to take place. Inputs include raw materials, labour, energy and capital.
- **Processes** are the industrial activities that take place in the factory to make the finished product. For example, in the car industry processes include moulding sheet steel into the shaped panels that make up the car, welding and painting.
- **Outputs** comprise the finished product or products that are sold to customers. Sometimes **by-products** and **waste products** may be produced.

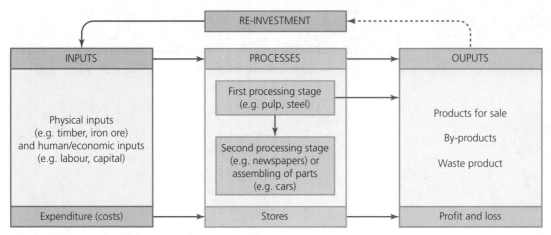

▲ **Figure 3.11** Industrial systems diagram

Manufacturing is often described or classified by the use of opposing terms, such as heavy industry and light industry (Table 3.9). In this case iron and steel would be an example of a heavy industry, using large amounts of bulky raw materials, processing on a huge scale and producing final products of a significant size. In contrast, the assembly of computers is a light industry.

Processing and assembly industries

- Processing industries are based on the direct processing of raw materials. Again, the iron and steel industry would be an example, using large quantities of iron ore, coal and limestone. Processing industries are often located close to their raw materials.

- Assembly industries put together parts and components that have been made elsewhere. A large car assembly plant will use thousands of components to build a car. Assembly industries usually have a much wider choice of location than processing industries and thus they are often described as '**footloose**' industries.

High-technology industry

- High-technology industry is the fastest growing manufacturing industry in the world. Virtually all developed countries and NICs have at least one high-technology cluster.

- 'High-tech' companies use or make silicon chips, computers, software, robots, aerospace components and other very technically advanced products. These companies put a great deal of money into scientific research. Their aim is to develop newer, even more advanced products.

- High-technology industries often cluster together in science parks. They are often found in close proximity to leading universities because of the need to employ well-qualified graduates in science and technology and to be aware of the latest research taking place in universities. The Cambridge Science Park is a major example in the UK.

▼ **Table 3.9** Classification of industry

Classification contrasts	Characteristics
Large scale and small scale	Depending on the size of plant and machinery, and the numbers employed.
Heavy and light	Depending on the nature of processes and products in terms of unit weight.
Market oriented and raw material oriented	Where the location of the industry is drawn either towards the market or the raw materials required.
Processing and assembly	Processing involves the direct processing of raw materials; assembly is to do with putting together parts and components.
Capital intensive and labour intensive	Depending on the ratio of investment in plant and machinery to the number of employees.
National and transnational	Many firms in the small- to medium-size range manufacture in only one country. Transnationals produce in at least two countries.

Factors affecting the location of industry

 PAGES 207–208

REVISED ☐

The factors affecting industrial location can be broadly subdivided into physical and human (Table 3.10). They relate both to individual factories and to industrial zones.

▼ **Table 3.10** Human and physical factors influencing industrial location

Physical factors	Human factors
Site: The availability and cost of land is important. Large factories will need flat, well-drained land on solid bedrock. An adjacent water supply may be essential.	**Capital (money)**: Some areas are more likely to attract investment than others.
Raw materials: Industries requiring heavy and bulky raw materials tend to locate as close as possible to these raw materials.	**Labour**: The quality and cost of labour are most important. The reputation, turnover, mobility and quantity of labour can also be important.
Energy: Energy-hungry industries, such as metal smelting, may be drawn to countries with relatively cheap hydro-electricity such as Norway.	**Transport and communications**: Transport costs remain important for heavy, bulky items. Accessibility to airports, ports, railway terminals and motorways may be crucial for some industries.
Natural routeways and harbours: Many modern roads and railways still follow natural routeways. Natural harbours provide good locations for ports and the industrial complexes often found at ports.	**Markets**: The location and size of markets are a major influence for some industries.
Climate: Some industries such as aerospace and film benefit directly from a sunny climate. Indirect benefits include lower heating bills and a more favourable quality of life.	**Government influence**: Government policies and decisions can have a big direct and indirect impact on the location of industry.
	Quality of life: Highly skilled personnel will favour areas where the quality of life is high.

The combined influence of a range of factors will impact on the decision-making of a company in terms of:

● Location – companies decide on particular locations for a variety of reasons. Most will look to the location that is seen as the 'least-cost location' or the 'highest-profit' location. A poor choice of location can mean a company making a loss and eventually closing.

● Scale of production – the amount of a product a company plans to produce will be an important factor in deciding location. Companies can achieve economies of scale by manufacturing more of a product.

- Methods of organisation – companies can follow various methods of organisation from traditional to highly innovative. Location factors can influence such decisions.

- The product or range of products manufactured – many large companies produce a range of products. Some locations may be more suited to the production of one product than another because of the cost factors involved.

Industrial agglomeration PAGES 208–210

REVISED

Industrial agglomeration can result in companies enjoying the benefits of external economies of scale. This means the lowering of a firm's costs due to external factors. External economies of scale can be subdivided into urbanisation economies and localisation economies.

Increasingly, industrial enterprises are located together on **industrial estates**. Industrial estates are usually located close to transport infrastructure, especially where more than one transport mode meet. The logic behind industrial estates includes:

- concentrating dedicated infrastructure in a small area to reduce the per-business expense of that infrastructure

- attracting new business by providing an integrated infrastructure in one location

- separating industry from residential areas to try to reduce the environmental impact

- eligibility of industrial estates for grants and loans under regional economic development policies.

The changing location of manufacturing

Changes in the location of manufacturing industry can be recognised at a range of scales:

- The global shift in manufacturing industry from the developed world to NICs and developing countries.

- Within individual countries the most significant locational change has been from traditional manufacturing regions, often on coalfields, to higher quality of life regions.

- Within individual regions of countries, manufacturing has historically concentrated in and around the major urban areas. However, in recent decades there has been a significant shift of industry towards rural **greenfield locations**.

- At the urban scale the relative shift from inner city to suburbs increased as the twentieth century progressed.

Common error

Error	Why it is wrong
'Using the word "industry" without specifying. The term can be applied to all sectors of the economy (for example, the agricultural industry and the service industry).'	If you use it with reference to the manufacture of goods then clearly state that this is 'manufacturing industry'.

Case study: Bangalore – India's high-tech city

- Bangalore is the most important city in India for high-technology industry. Known as the 'Garden City', Bangalore claims to have the highest quality of life in the country.
- In the 1980s Bangalore became the location for the first large-scale foreign investment in high technology in India when Texas Instruments selected the city.
- Other TNCs soon followed as the reputation of the city grew. Important backward and forward linkages were steadily established over time.
- Apart from ICT industries, Bangalore is also India's most important centre for aerospace and biotechnology.
- Many European and North American companies that previously outsourced their ICT requirements to local companies are now using Indian companies.
- The city's population grew from 2.4 million in 1981 to over 12 million in 2017. The city has grown into a major international hub for ICT companies.
- Bangalore has built up a large pool of highly-skilled labour. There has been very high investment into the city's infrastructure.

Exam-style questions

1 a What is the difference between processing and assembly industries? Give one example of each. [4]
  b Define high-technology industry. [2]
  c Discuss the factors that cause high-technology industries to cluster together. [3]
2 Discuss the reasons for the development of an industrial area you have studied. [6]

Answers on pages 132–133

3.4 Tourism

Key objectives

You should be able to:

- describe and explain the growth of tourism in relation to the main attractions of the physical and human landscape
- evaluate the benefits and disadvantages of tourism to receiving areas

- demonstrate an understanding that careful management of tourism is required in order for it to be sustainable.

Case study required

- An area where tourism is important.

Key definitions

Term	Definition
Tourism	Travel away from the home environment: (a) for leisure, recreation and holidays; (b) to visit friends and relations (VFR); (c) for business and professional reasons.
Package tour	The most popular form of foreign holiday where travel, accommodation and meals may all be included in the price and booked in advance.
Growth pole	A particular location where economic development, in this case tourism, is focused, setting off wider growth in the region as a whole.
Economic leakages	The part of the money a tourist pays for a foreign holiday that does not benefit the destination country because it goes elsewhere.
Multiplier effect	The idea that an initial amount of spending or investment causes money to circulate in the economy, bringing a series of economic benefits over time.
Sustainable tourism	Tourism organised in such a way that its level can be sustained in the future without creating irreparable environmental, social and economic damage to the receiving area.
Destination footprint	The environmental impact caused by an individual tourist.
Ecotourism	A specialised form of tourism where people experience relatively untouched natural environments, such as coral reefs, tropical forests and remote mountain areas, and ensures that their presence does no further damage to these environments.
Preservation	Maintaining a location exactly as it is and not allowing development.
Conservation	Allowing for developments that do not damage the character of a destination.
Community tourism	A form of tourism which aims to include and benefit local communities, particularly in developing countries.
Pro-poor tourism	Tourism that results in increased net benefits for poor people.

Over the last 50 years **tourism** has developed into a major global industry that is still expanding rapidly (Figure 3.12). It is one of the major elements in the process of globalisation.

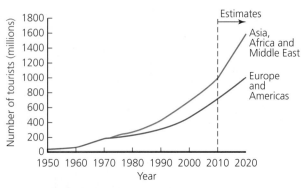

▲ **Figure 3.12** The growth in global tourism

The growth of tourism PAGES 211–214

- Tourism has developed in response to the main attractions of the physical and human landscape.
- During the seventeenth century, doctors increasingly began to recommend the benefits of mineral waters, and by the end of the eighteenth century there were hundreds of spas in Britain.
- The second stage in the development of holiday locations was the emergence of the seaside resort as people were beginning to appreciate coastal landscapes in a new way.
- The annual holiday was a product of the Industrial Revolution, which brought big social and economic changes. However, until the latter part of the nineteenth century only the very rich could afford to take a holiday away from home.
- The first **package tours** were arranged by Thomas Cook in 1841 in the UK. At the time it was the newly laid railway network that provided the transport infrastructure for Cook to expand his tour operations.
- Of equal importance was the emergence of a significant middle-class with time and money to spare for extended recreation. There was also a growing appreciation of what human landscapes such as major cities could offer.
- By far the greatest developments have occurred since the end of the Second World War.
- However, it took the jet plane to herald the era of international mass tourism.

Common error

Error	Why it is wrong
'Tourism only includes people on holiday.'	The definition of tourism also includes business and professional travel, and visits to friends and relations.

Reasons for the growth of global tourism

▼ **Table 3.11** Factors affecting global tourism

Economic	• Steadily rising real incomes and the decreasing real costs of holidays • Widening range of destinations within the middle-income range • Heavy marketing of shorter foreign holidays • Expansion of budget airlines • 'Globalisation' has increased business travel considerably
Social	• Increase in the average number of days of paid leave • Increasing desire to experience different cultures and landscapes • Raised expectations of international travel due to mass media • High levels of international migration means that more people have relatives and friends living abroad
Political	• Many governments have invested heavily to encourage tourism • Government backing for major international events such as the Olympic Games and the World Cup

Recent data

- In 2012 international tourist arrivals worldwide exceeded 1 billion for the first time ever. The WTO forecasts an increase to 1.8 billion in 2030.

- International tourism receipts reached $1260 billion in 2015. Tourism accounts for 7 per cent of the world's exports in goods and services and one in eleven jobs.
- Europe remains the world region with the greatest number of both tourist arrivals and tourism receipts, but many emerging economies have shown very fast growth rates in recent years.
- Leisure, recreation and holidays account for 53 per cent of inbound tourism. The second most important reason was visiting friends and relatives.
- Seasonality is the major problem with tourism as a source of employment.

The benefits and disadvantages of tourism to receiving areas 📖 PAGES 214–216

PAGES 214–216

REVISED

The economic impact

Supporters of the development potential of tourism say that:

- Tourism brings in valuable foreign currency and benefits other sectors of the economy.
- It provides considerable tax revenues.
- By providing employment in rural areas it can help to reduce rural–urban migration.
- A major tourism development can act as a **growth pole**.
- It can create openings for small businesses and support many jobs in the informal sector.

> **Tip**
>
> It is easy to fall into the trap of seeing only the advantages of the economic impact of tourism. It is always important to consider the other side of the coin, even if you can only come up with a few points.

▲ **Figure 3.13** The direct and indirect economic impact of the tourist industry

▲ **Figure 3.14** Economic leakages

However, critics say that the value of tourism is often overrated because:

- **economic leakages** (Figure 3.14) are high
- most local jobs created are menial, low-paid and seasonal
- money borrowed to invest in the necessary infrastructure for tourism increases the national debt
- at some destinations tourists spend most of their money in their hotels with minimum benefit to the wider community

> **Test yourself**
>
> 1 When did the number of international tourist arrivals first exceed 1 billion?
>
> 2 What is a *growth pole*?
>
> 3 Define economic leakages.
>
> **Answers on page 126**

- tourism might not be the best use for local resources that could in the future create a larger **multiplier effect** if used by a different economic sector
- locations can become over-dependent on tourism
- the tourist industry has a huge appetite for resources.

The social and cultural impact

▼ **Table 3.12** Disadvantages and advantages of the development of tourism

Social/cultural disadvantages	Social/cultural advantages
• Loss of locally owned land • Abandonment of traditional values • Displacement of people • Traditional community structures may be weakened • Abuse of human rights • Increasing availability of alcohol and drugs • Crime and prostitution, sometimes involving children • Visitor congestion at key locations • Denying local people access to beaches • Loss of housing for local people as more visitors buy second homes	• Can increase the range of social facilities for local people • Can lead to greater understanding between people of different cultures • Visiting ancient sites can develop a greater appreciation of the historical legacy of host countries • Can help develop foreign language skills in host communities • May encourage migration to major tourist generating countries • Major international events such as the Olympic Games can have a very positive global impact

The management and sustainability of tourism

 PAGES 217–220

REVISED

Tourism has reached such a large scale in so many parts of the world that it can only continue with careful management. However, **sustainable tourism** strategies have been much more successful in some areas than others.

Environmental groups are keen to make tourists aware of their **'destination footprint'**. This is the environmental impact caused by an individual tourist. People are being urged to:

- 'fly less and stay longer'
- carbon-offset their flights
- consider 'slow travel'.

Virtually every aspect of the industry now recognises that tourism must become more sustainable. **Ecotourism** is at the leading edge of this movement.

Protected areas

Over time, more and more of the world's most spectacular and ecologically sensitive areas have been designated for protection. The world's first National Park was established at Yellowstone in the USA in 1872. Now there are well over 1000 worldwide. Many countries have National Forests, Country Parks, Areas of Outstanding Natural Beauty, World Heritage Sites and other designated areas that merit special status and protection.

In many countries and regions there are often differences of opinion when the issue of special protection is raised. For example, in some areas jobs in mining, forestry and tourism may depend on developing presently unspoilt areas. Often, a clear distinction has to be made between the objectives of **preservation** and **conservation**.

Tourist hubs

The idea is to concentrate tourism and its impact in one particular area so that the majority of the region or country feels little of the negative impacts of the industry. Benidorm in Spain and Cancun in Mexico are examples where the model was adopted but both locations show how difficult it is to confine tourism within preconceived boundaries as the number of visitors increases and people want to travel beyond tourist enclaves.

Quotas

Quotas set limits on the number of people visiting a location. This is an idea we are likely to hear much more about in the future.

Case study: Jamaica – the benefits and disadvantages associated with the growth of tourism

- Tourism has become an increasingly vital part of Jamaica's economy in recent decades.
- Tourism's direct and indirect contribution to GDP in 2014 amounted to almost 27.2 per cent of total GDP.
- Direct employment in the industry amounted to 90 000.
- Tourism is the largest source of foreign exchange for the country.
- The Jamaican government sees the designation of the National and Marine Parks as a positive environmental impact of tourism. Entry fees to the Parks pay for conservation.
- The Marine Parks are attempting to conserve the coral reef environments off the coast of Jamaica.
- Ecotourism is a developing sector of the industry.
- Considerable efforts are being made to promote **community tourism**, which is seen as an important aspect of '**pro-poor tourism**'.
- The physical attractions of Jamaica almost sell themselves, so the government is putting much effort into trying to boost the island's human attractions.
- During the off-season, 25 per cent of hotel workers are laid off.
- Other negative aspects include: the environmental impact of tourism; the heavy use of resources, particularly water, by hotels; socio-cultural problems between residents and visitors.

Exam-style questions

1 a Define tourism.   [2]
   b Discuss three reasons for the growth of international tourism.   [3]
   c Examine the major social issues associated with the
      development of tourism.   [4]
2 a Why is it important to be aware of the carrying capacity of a
      tourist destination?   [3]
   b With reference to an example, explain the meaning of
      ecotourism.   [4]

Answers on page 133

3.5 Energy

Key definitions

REVISED

Term	Definition
Fossil fuels	Fuels consisting of hydrocarbons (coal, oil and natural gas), formed by the decomposition of prehistoric organisms in past geological periods.
Renewable energy	Sources of energy such as solar and wind power that are not depleted as they are used.
Energy mix	The relative contribution of different energy sources to a country's energy consumption.
Biofuels	Fossil fuel substitutes that can be made from a range of crops including oilseeds, wheat and sugar. They can be blended with petrol and diesel.
Geothermal energy	The natural heat found in the Earth's crust in the form of steam, hot water and hot rock.

Non-renewable and renewable energy supplies PAGES 221–224

REVISED

Non-renewable sources of energy are the **fossil fuels** and nuclear fuel. Eventually, these non-renewable resources could become completely exhausted. The burning of fossil fuels creates pollution and is the major source of greenhouse gas emissions. Climate change due to these emissions is the biggest environmental problem facing the planet.

Renewable energy resources are mainly forces of nature that are sustainable and which usually cause little or no pollution. Renewable energy includes hydro-electricity, biofuels, and wind, solar, geothermal, tidal and wave power.

At present, non-renewable resources dominate global energy. The challenge is to transform the global **energy mix** to achieve a better balance between renewables and non-renewables. There is a huge gap in energy consumption between rich and poor countries. Wealth is the main factor explaining the energy gap.

The demand for energy has grown steadily over time with a global increase of over 60 per cent between 1990 and 2015 (Figure 3.15). The fossil fuels dominate the global energy situation. Their relative contribution in 2015 was: oil 37 per cent, coal 17 per cent and natural gas 31 per cent. In contrast, hydro-electricity accounted for 2.5 per cent and nuclear energy 8.3 per cent. Consumption by type of fuel varies widely by world region.

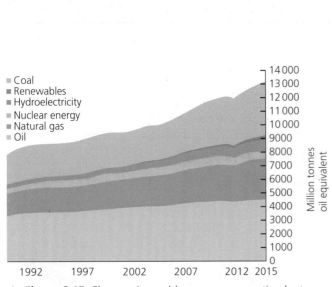

▲ **Figure 3.15** Changes in world energy consumption by type, 1990–2015

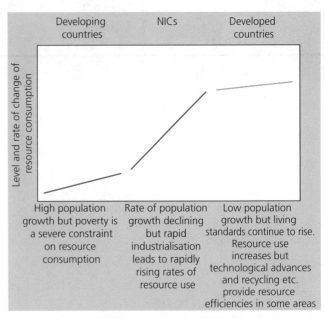

▲ **Figure 3.16** Model of the relationship between resource use and the level of economic development

The highest energy consumers, such as the USA and Canada, use more than 6 tonnes oil equivalent per person, while almost all of Africa and much of South America and Asia use less than 1.5 tonnes oil equivalent per person. Figure 3.16 is a model showing the relationship between resource use in general and the level of economic development. This model applies well to energy consumption.

Common error

REVISED

Error	Why it is wrong
'Production and consumption are the same thing.'	For some energy sources such as coal the figures are very similar, but for oil there is a very significant difference. The ease with which a type of energy can be transported is a major factor here.

Fuelwood in developing countries

- In developing countries about 2.5 billion people rely on fuelwood as their main source of energy.
- Fuelwood provides much of the energy needs for sub-Saharan Africa. It is also the most important use of wood in Asia.
- In 2012, 1.1 billion people were still living without electricity.
- The transition from fuelwood and animal dung to 'higher level' sources of energy (the energy ladder) occurs as part of the economic development process.

The benefits and disadvantages of nuclear power and renewable energy 📖 PAGES 224–230

Nuclear power

The main concerns about nuclear power are:

- Power plant accidents, which could release radiation into air, land and sea.
- Radioactive waste storage/disposal. Most concern is over the small proportion of 'high-level waste'.
- Rogue state or terrorist use of nuclear fuel for weapons.
- High construction and decommissioning costs.
- Seen by some people as less 'democratic' than other sources of power.
- The possible increase in certain types of cancer near nuclear plants.

The advantages of nuclear power are:

- Zero emissions of greenhouse gases.
- Reduced reliance on imported fossil fuels.
- Not as vulnerable to fuel price fluctuations as oil and gas.
- In recent years nuclear plants have demonstrated a very high level of reliability.
- Nuclear technology has spin-offs in fields such as medicine and agriculture.

Renewable energy supplies

Countries are eager to harness renewable energy resources to:

- reduce their reliance on domestic fossil fuel resources
- lower their reliance on costly fossil fuel imports
- improve their energy security
- cut greenhouse gas emissions.

Hydro-electricity

HEP is by far the most important source of renewable energy. The 'big four' HEP nations of China, Brazil, Canada and the USA account for almost 54 per cent of the global total. Most of the best HEP locations are already in use so the scope for more large-scale development is limited.

Although HEP is generally seen as a clean form of energy, it is not without its problems, which include:

- large dams and power plants can have a huge negative visual impact on the environment
- the obstruction of the river for aquatic life
- deterioration in water quality
- large areas of land may need to be flooded to form the reservoir behind the dam
- submerging large forests without prior clearance can release significant quantities of methane, a greenhouse gas.

Newer alternative energy sources

The main drawback to the new alternative energy sources is that they usually produce higher-cost electricity than traditional sources. However, the cost gap with non-renewable energy is narrowing. Figure 3.17 shows the sharp increase in the consumption of renewable energy (other than HEP) in the last decade. In 2015 this accounted for almost 2.8 per cent of global primary energy consumption. The newer sources of renewable energy making the largest contribution to global energy supply are wind power and biofuels.

Wind power

The worldwide capacity of wind energy reached almost 432 GW by the end of 2015. The leaders in global wind energy are China, the USA, Germany, India and Spain. Together these countries account for over 67 per cent of the world total. The main advantages of wind energy are that, compared with most other forms of renewable energy, it can generate significant amounts of electricity and it can be harnessed to a reasonable degree in most parts of the world.

As wind turbines have been erected in more areas of more countries, the opposition to this form of renewable energy has increased:

- People are concerned that huge turbines located nearby could blight their homes.
- There are concerns about the hum of turbines disturbing both people and wildlife.
- Skylines in scenically beautiful areas might be spoiled forever.
- Turbines can kill birds.
- Suitable areas for wind farms are often near the coast where land is both scenically beautiful and expensive.

The development of large offshore wind farms has become an increasingly debatable issue. There has also been increasing debate about how much electricity wind turbines in many areas actually produce. There can be a big difference between the technical capacity of a wind turbine and the amount of electricity it actually produces.

Biofuels

The biggest producers of biofuels are the USA, Brazil and Germany. Advocates of biofuels argue that biofuels come from a renewable resource (crops); can be produced wherever there is sufficient crop growth, helping energy security; and often produce cleaner emissions than petroleum-based fuels.

However, there are clear disadvantages in biofuel production. Increasing amounts of cropland have been used to produce biofuels, adding to the 'global food crisis'. Large amounts of land, water and fertilisers are needed for large-scale crop production. The manufacture of biofuels also uses significant amounts of energy, creating greenhouse gas emissions. In addition, biofuels have a lower energy output than traditional fuels.

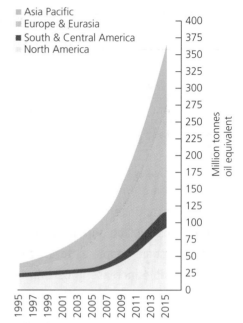

▲ **Figure 3.17** Renewable energy consumption by world region, 1995–2015

Geothermal energy

Geothermal energy can be used directly for industry, agriculture, bathing and cleansing. For example, in Iceland, hot springs supply water at 86 °C to 95 per cent of the buildings in and around Reykjavik.

The USA is the world leader in geothermal electricity. At present virtually all the geothermal power plants in the world operate on steam resources, and have an extremely low environmental impact.

The advantages of geothermal power are:

- extremely low environmental impact
- geothermal plants occupy relatively small land areas
- not dependent on weather conditions (like wind and solar power)
- relatively low maintenance costs.

The limitations of this form of energy are:

- there are few locations worldwide where significant amounts of energy can be generated
- total global generation remains very small
- some of these locations are far from where the energy could be used
- installation costs of plant and piping are relatively high.

Solar power

In 2016 global solar power capacity reached 305 GW. China, Japan, Germany, the USA and Italy currently lead the global market for solar power.

Solar electricity is currently produced in two ways:

- Photovoltaic systems – these are solar panels that convert sunlight directly into electricity.
- Concentrating solar power (CSP) systems use mirrors or lenses and tracking systems to focus a large area of sunlight into a small beam. This concentrated light is then used as a heat source for a conventional thermal power plant.

▼ **Table 3.13** Advantages and disadvantages of solar power

Advantages	Disadvantages
A completely renewable resource	Initial high cost of solar plants
No noise or direct pollution	Solar power cannot be harnessed during storms, on cloudy days or at night
Very limited maintenance required	Of limited use in countries with low annual hours of sunshine
Technology is improving and reducing costs	Large areas of land required to capture the Sun's energy in order to generate significant amounts of power
Can be used in remote areas where it is too expensive to extend the electricity grid	
A generally positive public perception	

Tidal and wave power

Tidal power plants act like underwater windmills, transforming sea currents into electrical current. Tidal power is more predictable than solar or wind power, and the infrastructure is less obtrusive. However, start-up costs are high. While generally predictable, tidal energy is not as dependable as fossil-fired or nuclear generation.

Wave energy is where generators are placed on the ocean's surface and energy levels are determined by the strength of the waves. The costs and benefits of wave energy are broadly similar to those of tidal power.

The development of both of these sources of energy is at a very early stage.

> **Tip**
>
> Solar power is generally taken to mean the production of solar electricity, as distinct from solar hot water systems.

Test yourself

1 What is *renewable energy*?
2 How many people in developing countries rely on fuelwood as their main source of energy?
3 Name the leading countries in global wind energy production.

Answers on page 127

Case study: Energy supply in China

- China uses more energy than any other country in the world. In 2015 China's main sources of energy were: coal (63.7 per cent), oil (18.6 per cent) and hydro-electricity (8.5 per cent).

- China was an exporter of oil until the early 1990s although it is now a very significant importer.

- Chinese investment in energy resources abroad has risen rapidly in order to achieve long-term energy security.

- In recent years China has tried to take a more balanced approach to energy supply and at the same time sought to reduce its environmental impact. The development of clean coal technology is an important aspect of this approach.

- The further development of nuclear and hydropower is another important strand of Chinese policy.

- China aims to increase the production of oil while augmenting that of natural gas and improving the national oil and gas network. Priority has also been given to building up the national oil reserve.

- Total renewable energy capacity in China reached 502 GW in 2015. This included 319 GW of hydro-electricity, 129 GW of wind energy, 43 GW of solar PV and 10 GW of bioenergy.

- The Three Gorges Dam across the Yangtze river is the world's largest electricity generating plant of any kind. This is a major part of China's policy in reducing its reliance on coal.

Sample exam questions

REVISED

a Define renewable energy. [2]

Student's answer

> a Renewable energy can be used over and over again. These resources are mainly forces of nature that are sustainable and which usually cause little or no environmental pollution. Examples are wind and solar power.

Teacher's comments

A good, clear definition with two examples provided. Full marks.

b Why is fuelwood such an important source of energy in the
developing world?   [3]

Student's answer

b In developing countries about 2.5 billion people rely on fuelwood, charcoal
and animal dung for cooking. Fuelwood and charcoal are collectively called
fuelwood, which accounts for just over half of global wood production. Fuelwood
provides much of the energy needs for sub-Saharan Africa. It is also the most
important use of wood in Asia. So many people rely on fuelwood because other
sources of energy are either not available where they live or they cannot afford
to pay for them.

Teacher's comments

A very good answer that (a) shows how many people are reliant on fuelwood
worldwide, (b) accurately defines fuelwood, and (c) states why so many
people do not have access to other forms of energy. The student gains all
three marks here.

c Discuss the advantages and disadvantages of nuclear power.   [6]

Student's answer

c There are many disadvantages of nuclear power. A nuclear power plant accident
could release radiation into the atmosphere. There are big concerns about the
storage of nuclear waste, particularly high-level waste. Nuclear power stations
cost a great deal of money not just to build, but also to decommission when
they can no longer produce energy effectively. There are also big security
concerns about nuclear power. An advantage of nuclear power is that it does
not produce greenhouse gases.

Teacher's comments

This is a good answer with regard to disadvantages with four significant
concerns identified. However, only one advantage is considered. Because
of this lack of balance the student only achieves four marks out of the six
available. Other advantages that could be considered include: (a) reduced
reliance on imported fossil fuels, (b) the increasing efficiency and reliability
of nuclear energy, and (c) the fact that nuclear power is not as vulnerable to
fuel price fluctuations as oil and gas.

Exam-style questions

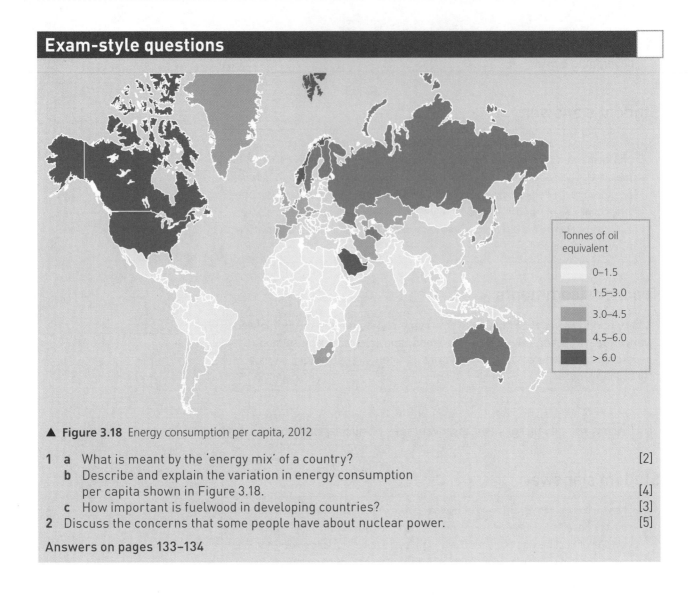

▲ **Figure 3.18** Energy consumption per capita, 2012

1 a What is meant by the 'energy mix' of a country?   [2]
b Describe and explain the variation in energy consumption
per capita shown in Figure 3.18.   [4]
c How important is fuelwood in developing countries?   [3]
2 Discuss the concerns that some people have about nuclear power.   [5]

Answers on pages 133–134

Key objectives

You should be able to:

- describe methods of water supply and the proportions of water used for agricultural, domestic and industrial purposes in countries at different levels of economic development

- explain why there are water shortages in some areas and demonstrate that careful management is required to ensure future supplies.

Case study required

- Water supply in a country or area.

Key definitions

REVISED

Term	Definition
Water supply	The provision of water by public utilities, commercial organisations or by community endeavours.
Dam	A barrier that holds back water.
Reservoir	An artificial lake primarily used for storing water.
Wells and boreholes	A means of tapping into various types of aquifers (water-bearing rocks), gaining access to groundwater.
Potable water	Water that is free from impurities, pollution and bacteria, and is thus safe to drink.
Water stress	When water supply is below 1700 cubic metres per person per year.

The global water crisis PAGE 231

REVISED

- For about 80 countries, with 40 per cent of the world's population, lack of water is a constant threat.
- Demand for water is doubling every 20 years.
- In the poorest nations water supplies are often polluted.

Methods of water supply PAGES 231–234

REVISED

The objective in all methods of water supply is to take water from its source to the point of usage. In 2015 about 91 per cent of the global population had access to piped water supply through house connections or an improved water source (including through standpipes). This left over 660 million people who did not have access to an improved water source.

Dams and reservoirs

In the twentieth century, global water consumption grew sixfold, twice the rate of population growth. Much of this increased consumption was made possible by investment in water infrastructure, particularly dams and reservoirs, affecting nearly 60 per cent of the world's major river basins.

Dams are mainly used to save, manage and prevent the flow of excess water into specific regions. Not all reservoirs are held behind dams, but the really large ones usually are. These are 'on channel' reservoirs where a dam has been built across an existing river. In contrast, 'off channel' reservoirs usually use depressions in the existing landscape or human-dug depressions to store water.

Globally, the construction of dams has declined since the 1960s and 1970s. This is because most of the best sites for dams are already in use or such sites are strongly protected. An alternative to building new dams and reservoirs is to increase the capacity of existing reservoirs by extending the height of the dam.

Wells and boreholes

Wells and boreholes are sunk directly down to the water table. The water table is the highest level of underground water. For many communities groundwater is the only water supply source. Aquifers provide approximately half of the world's drinking water, 40 per cent of the water used by industry and up to 30 per cent of irrigation water. Typically, a borehole is drilled by machine and is relatively small in diameter. Wells are relatively large in diameter and are often sunk by hand although machinery may be used.

About 35 per cent of all public water supply in England and Wales comes from groundwater. Groundwater is even more important in arid and semi-arid areas. This is the main source of water of oasis settlements such as those in the Sahara desert in North Africa.

Desalination: the answer to water shortages?

Desalination plants are in widespread use in the Middle East where other forms of water supply are extremely scarce. Most of these plants distil water by boiling, generally using waste gases produced by oil wells. Without the availability of waste energy the process would be extremely expensive. This is the main reason why desalination plants are few and far between outside of the Middle East.

Other methods of water supply

There are a number of other methods of water supply, shown in Figure 3.88 on page 232 of the textbook, which are largely self-explanatory.

How water use varies PAGES 234–235 REVISED

Figure 3.19 contrasts water use in developed and developing countries. In the latter, agriculture accounts for over 80 per cent of total water use. Large variations in water allocation can also exist within countries. For example, irrigation accounts for over 80 per cent of water demand in the west of the USA, but only about 6 per cent in the east. In general, precipitation declines from east to west in the USA.

The amount of water used by a population depends not only on water availability but also on levels of urbanisation and economic development. As global urbanisation continues, the demand for **potable water** in cities and towns will rise rapidly.

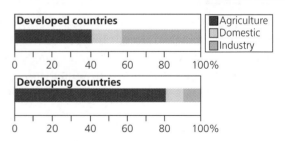

▲ **Figure 3.19** Water used for agriculture, industry and domestic purposes in developed and developing countries

Water shortages PAGES 231–236 REVISED

Much of the precipitation that falls to the Earth's surface cannot be captured and the rest is very unevenly distributed. For example:

- Over 60 per cent of the world's population live in areas receiving only 25 per cent of global annual precipitation.
- The arid regions of the world cover 40 per cent of the world's land area, but receive only 2 per cent of global precipitation.

Water scarcity is to do with the availability of potable water:

- **Physical water scarcity** is when physical access to water is limited. This is when demand outstrips a region's ability to provide the water needed by the population. It is the arid and semi-arid regions of the world that are most associated with physical water scarcity.
- **Economic water scarcity** exists when a population does not have the necessary monetary means to utilise an adequate source of water. The crux of the problem is lack of investment. Much of sub-Saharan Africa is affected by this type of water scarcity.

Securing access to clean water is a vital aspect of development. While deaths associated with dirty water have been virtually eliminated from developed countries, in developing countries most deaths still result from water-borne disease.

Water scarcity has been presented as the 'sleeping tiger' of the world's environmental problems, threatening to put world food supplies in jeopardy, limit economic and social development, and create serious conflicts between neighbouring drainage basin countries. The UN estimates that by 2025, 1.8 billion people are expected to be living in countries or regions with absolute water scarcity, and two-thirds of the world population could be under water stress conditions.

A country is judged to experience **water stress** when water supply is below 1700 cubic metres per person per year. When water supply falls below 1000 cubic metres per person a year, a country faces water scarcity for all or part of the year.

The link between poverty and water resources is very clear, with those living on less than $1.25 a day roughly equal to the number without access to safe drinking water.

The future

Scientists expect water scarcity to become more severe because:

- the world's population continues to increase significantly
- increasing affluence is inflating per capita demand for water
- of increasing demands of biofuel production – biofuel crops are heavy users of water
- climate change is increasing aridity and reducing supply in many regions
- many water sources are threatened by various forms of pollution.

Test yourself

1 How many people do not have access to an improved water source?

2 What is the process that can change salt water into potable water?

3 Define *water scarcity*.

Answers on page 127

Water management  PAGES 237–240

Many experts say water should cost more for users. This would encourage more careful use of water. However, many consumers still see water as a 'free' or very low-cost resource and there is concern that higher prices would impact unfairly on people on low incomes.

Conserving irrigation water would have more impact than any other measure. Most irrigation is extremely inefficient, wasting half or more of the water used. Although some industries have significantly reduced their use of water per unit of production, most water analysts believe that much more can be done. Urban sanitation services are very heavy users of water. Demand could be reduced considerably by adopting dry, or low water use, systems.

As water scarcity becomes more of a problem, the investment required to tackle this global challenge will have to rise significantly.

Case study: The water problem in southwestern USA

- The USA is a huge user of water. The western states of the USA, covering 60 per cent of the land area with 40 per cent of the total population, receive only 25 per cent of the country's mean annual precipitation. Yet each day the west uses as much water as the east.

- The southwest in particular has prospered due to a huge investment in water transfer schemes. This has benefited agriculture, industry and settlement.

- California has benefited most from this investment in water supply. Seventy per cent of runoff originates in the northern one-third of the state but 80 per cent of the demand for water is in the southern two-thirds. While irrigation is the prime water user, the sprawling urban areas have also greatly increased demand.

- The 2333-km long Colorado river is an important source of water in the southwest. Over 30 million people in the region depend on water from the river. Despite the interstate and international agreements (between the USA and Mexico), major problems over the river's resources have arisen because population has increased along with rising demand from agriculture and industry.

- The $4 billion Central Arizona Project (CAP) is the latest, and probably the last, big money scheme to divert water from this great river.

- Resource management strategies include: measures to reduce leakage and evaporation losses; recycling more water in industry; charging more realistic prices for irrigation water; extending the use of the most efficient irrigation systems; changing from highly water-dependent crops such as rice and alfalfa to those needing less water.

- Future options include: developing new groundwater resources; investing in more desalination plants; constructing offshore aqueducts that would run under the ocean from the Columbia river in the northwest of the USA to California.

- There is now general agreement that planning for the future water supply of the southwest should embrace all practicable options.

Exam-style questions

1 a Why do some water experts talk about a 'global water crisis'? [3]
  b Define the term water supply. [2]
  c How important are dams and reservoirs to global water supply? [3]
2 a How are wells and boreholes used to provide water supply? [2]
  b Discuss two other methods of water supply illustrated in Figure 3.88 on page 232 of the textbook. [4]

Answers on page 134

3.7 Environmental risks of economic development

Key objectives

You should be able to:

- describe how economic activities may pose threats to the natural environment, locally and globally
- demonstrate the need for sustainable development and management
- understand the importance of resource conservation.

Case study required

- An area where economic development is taking place, causing the environment to be at risk.

Key definitions

Term	Definition
Pollution	Contamination of the environment. It can take many forms – air, water, soil, noise, visual and others.
Prevailing winds	The major direction of winds in a region.
Externalities	The side effects, positive and negative, of an economic activity that are experienced beyond its site.
Enhanced greenhouse effect	Large-scale pollution of the atmosphere by economic activities has created an enhanced greenhouse effect.
Deforestation	The loss of forested lands for a number of reasons including the clearing of land for agricultural use, for timber, and for other activities such as mining.
Overgrazing	The grazing of natural pastures at stocking intensities above the livestock carrying capacity.
Desertification	The gradual transformation of habitable land into desert.
Dust storm	A severe windstorm that sweeps clouds of dust across an extensive area, especially in an arid region.
Resource management	The control of the exploitation and use of resources in relation to environmental and economic costs.
Sustainable development	A carefully calculated system of resource management which ensures that the current level of exploitation does not compromise the ability of future generations to meet their own needs.
Environmental impact statement	A document required by law detailing all the impacts on the environment of an energy or other project above a certain size.
Conservation of resources	The management of the human use of natural resources to provide the maximum benefit to current generations while maintaining capacity to meet the needs of future generations.
Recycling	The concentration of used or waste materials, their reprocessing, and their subsequent use in place of new materials.
Reuse	Extending the life of a product beyond what was the norm in the past, or putting a product to a new use and extending its life in this way.
Quotas	Involving agreement between countries to take only a predetermined amount of a resource.
Rationing	A last resort management strategy when demand is massively out of proportion to supply. For example, individuals might only be allowed a very small amount of fuel and food per week.

Term	Definition
Subsidies	Financial aid supplied by the government to an industry for reasons of public welfare.
Carbon credit	A permit that allows an organisation to emit a specified amount of greenhouse gases.
Carbon trading	A company that does not use up the level of emissions it is entitled to can sell the remainder of its entitlement to another company.
Community energy	Energy produced close to the point of consumption.
Microgeneration	Generators producing electricity with an output of less than 50 KW.

The threat of economic activities to the natural environment 📖 PAGES 241–245

REVISED

As the scale of global economic activity has increased, the strain on the natural environment has become more obvious at all scales. The planet is experiencing a range of serious environmental challenges, many of which are interlinked.

Pollution

Pollution has a major impact on people and the environment. The methods of exposure to pollutants are inhalation, ingestion and absorption.

Air pollution

Of all types of pollution, air pollution has the most widespread effects on human health and the environment. In many parts of the developing world indoor air pollution is more severe than that experienced outdoors. This is the result of the use of biomass fuels for cooking and heating.

The most serious polluters are the large-scale processing industries which tend to form agglomerations. The impact of large industrial agglomerations may spread well beyond the locality to cross international borders. For example, **prevailing winds** in Europe generally carry pollution from west to east.

Air pollution is a massive environmental problem leading to, among other things, global warming, acid rain and the deterioration of the ozone layer. The major air pollutants include ozone, carbon monoxide, nitrogen dioxide, particulate matter, sulfur dioxide and lead.

Pollution is the major **externality** of industrial and urban areas. Pollution is at its most intense at the focus of pollution-causing activities, declining with distance from such concentrations. For some sources of pollution it is possible to map the externality gradient and field (Figure 3.20).

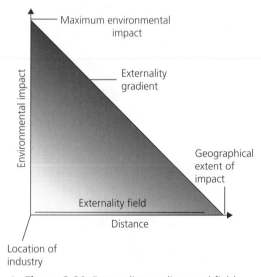

▲ **Figure 3.20** Externality gradient and field

Water pollution

Each year about 450 km³ of wastewater are discharged into rivers, streams and lakes around the world. While rivers in more affluent countries have become steadily cleaner, the reverse has been true in much of the developing world. It has been estimated that 90 per cent of sewage in developing countries is discharged into rivers, lakes and seas without any treatment. China's rapid economic growth has led to widespread environmental problems. Pollution problems are so severe in some areas that the term 'cancer villages' has become commonplace. The Chinese government admits that 300 million people drink polluted water.

Noise and light pollution

Noise and light pollution are increasing hazards in developed societies. Noise pollution is disturbing or excessive noise that may harm the activity or balance of human or animal life. Most outdoor noise is caused by machines and modes of transport. The increase in air traffic is one of the major contributors to noise pollution (and air pollution).

Sources of light pollution include the interior and external lighting of all sorts of buildings, advertising and street lighting. It is most severe in highly industrialised and densely populated areas. It can impact on human health, causing fatigue, loss of sleep, headaches and loss of amenity.

The relative risks of incidental and sustained pollution

It is important to consider the different impact between incidental pollution (one-off pollution incidents) and sustained pollution (longer-term pollution). The former is mainly linked to major accidents caused by technological failures and human error. Causes of the latter include ozone depletion and global warming. Two of the worst examples of incidental pollution have been Bhopal and Chernobyl.

Acid deposition

Acid deposition is a result of the mix of air pollutants, and leads to the acidification of freshwater bodies and soils. There are two forms:

- Dry deposition is in the form of particles, aerosols and gases, and occurs in the absence of precipitation.
- Wet deposition involves acid rain, snow, fog and mist. Acid rain is rainwater with a pH of less than 5.

Most pollutants causing acid rain result from human activity. This is mainly a result of the conversion of sulfur dioxide and oxides of nitrogen (NOx), produced when fossil fuels are burned, into the sulfates and nitrates of dry deposition and the sulfuric and nitric acids of wet deposition. Dry deposition usually occurs close to emission sources. Wet deposition can be carried great distances. The environmental effects include:

- damage to forests
- depletion of essential nutrients in soils
- increase in acidity of freshwater bodies.

There are two strategies for managing acid deposition:

- preventing its occurrence
- repairing the damage.

Test yourself

1 Draw a labelled diagram showing an externality gradient and field.
2 How much wastewater is discharged into water bodies each year?
3 What are the **two** strategies used to manage acid deposition?

Answers on page 127

Enhanced global warming  PAGES 248–250

Economic activities generating greenhouse gases

The Earth–atmosphere system has a natural greenhouse effect (Figure 3.21) that is essential to all life on Earth. However, large-scale pollution of the atmosphere by economic activities has created an **enhanced greenhouse effect**. This is causing temperatures to increase beyond the limits of the natural greenhouse effect. Many parts of the world are experiencing changes in their weather that are unexpected. Some of these changes could have disastrous consequences for the populations of the areas affected if they continue to get more severe.

The present rate of change is greater than anything that has happened in the past. In the twentieth century, average global temperatures rose by 0.6 °C. The predictions are for a further global average temperature increase of between 1.6 °C and 4.2 °C by 2100.

The main greenhouse gases being created by human activity are:

- carbon dioxide
- methane
- nitrous oxides
- chlorofluorocarbons
- ozone.

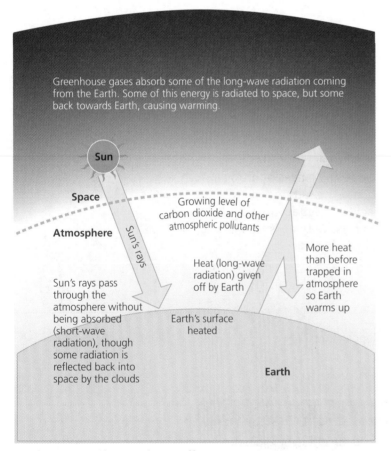

▲ **Figure 3.21** The greenhouse effect

The potential consequences of enhanced global warming include:

- greater global temperature variations and heatwaves
- rising sea levels
- increasing acidity in oceans
- melting of ice caps and glaciers
- disruption of the warm water currents of the Gulf Stream
- growth of the tropical belt
- changing patterns of rainfall
- declining crop yields.

Soil erosion and desertification PAGES 250–253 REVISED

The extent and causes of soil erosion

Soil degradation involves both the erosion and the reduction in quality of topsoil. The loss of the upper soil horizons containing organic matter and nutrients and the thinning of soil profiles reduce crop yields.

- Globally about 2 billion hectares of soil resources have been degraded. This is equivalent to about 15 per cent of the Earth's land area.
- During the past 40 years nearly one-third of the world's cropland has been abandoned because of soil erosion.

In temperate areas much soil degradation is a result of market forces and the attitudes adopted by commercial farmers and governments. In contrast, in the tropics much degradation results from high population pressure, land shortages and lack of awareness. The greater climate extremes and poorer soil structures in tropical areas also give greater potential for degradation.

The main cause of soil degradation is the removal of the natural vegetation cover, leaving the surface exposed to the elements. Figure 3.22 shows the human causes of degradation.

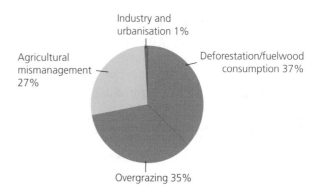

▲ **Figure 3.22** Causes of land degradation

Deforestation occurs for a number of reasons including the clearing of land for agricultural use. Such activities tend to happen quickly whereas the loss of vegetation for fuelwood, a massive problem in many developing countries, is generally a more gradual process. Population pressure in many areas and poor agricultural practices have resulted in serious **overgrazing**.

Figure 3.111 on page 251 of the textbook illustrates how a combination of causes and processes can operate in an area to result in soil degradation. The diagram shows a range of different economic activities which impact on the soil.

Environmental and socio-economic consequences of soil erosion

The environmental and socio-economic consequences of soil degradation are considerable (see Table 3.14 on page 108). **Desertification** is arguably the most serious environmental consequence of soil degradation. Desertification is usually caused by climate change and/or by destructive use of the land. Natural causes of desertification include temporary drought periods of high magnitude and long-term climate change towards aridity. The main human causes are overgrazing, overcultivation and deforestation.

Desertification occurs when already fragile land in arid and semi-arid areas is over-exploited. It is a considerable problem in many parts of the world, for example, on the margins of the Sahara desert in North Africa and the Kalahari desert in southern Africa. **Dust storms**, which can seriously damage crops, may also be a problem in such areas.

▼ **Table 3.14** The consequences of desertification

Environmental	Economic	Social and cultural
• Loss of soil nutrients through wind and water erosion • Changes in composition of vegetation and loss of biodiversity as vegetation is removed • Reduction in land available for cropping and pasture • Increased sedimentation of streams because of soil erosion and sediment accumulations in reservoirs	• Reduced income from traditional economy (pastoralism and cultivation of food crops) • Decreased availability of fuelwood, necessitating purchase of oil/ kerosene • Increased dependence on food aid • Increased rural poverty	• Loss of traditional knowledge and skills • Forced migration due to food scarcity • Social tensions in reception areas for migrants

The increasing world population and the changing diets of hundreds of millions of people as they become more affluent is placing more and more pressure on land resources. Some soil and agricultural experts say that a decline in long-term soil productivity is already seriously limiting food production in the developing world.

Test yourself

4 List **three** consequences of enhanced global warming.

5 What are the **two** major causes of land degradation?

6 Define *resource management*.

Answers on page 127

Sustainable development and management

 PAGES 254–256

REVISED

Resource management and **sustainable development** are vital for the future of the planet.

Figure 3.23 shows what has happened in so many of the world's fishing grounds. Without careful resource management fish stocks could be totally depleted in some areas. Yet, it is often difficult to get countries to agree on what to do. The European Union's Common Fisheries Policy is an international attempt to manage the fishing grounds belonging to this group of countries. The European Union also tries to manage agriculture in its member countries through its Common Agricultural Policy (CAP).

Environmental impact statements and pollution control

Most countries now require some form of **environmental impact statement** for major projects such as a new road, an airport or a large factory. The objective is to identify all the environmental consequences and to try to minimise these as far as possible.

Industry has spent increasing amounts on research and development to reduce pollution – the so-called 'greening of industry'. In general, after a certain stage of economic development the level of pollution will decline. This is because countries have become more aware of their environmental problems with higher levels of economic activity and they have also created the wealth to invest in improving the environment. Increasingly, successful policies developed in one country are being followed elsewhere.

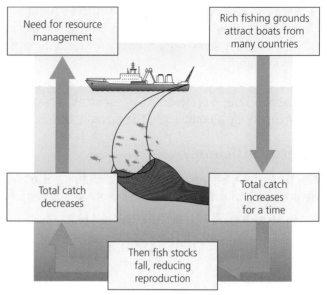

▲ **Figure 3.23** Fishing and resource management

The importance of resource conservation

 PAGES 256–259

Conservation includes both the protection and rational use of natural resources. Both the demand for, and supply of, resources need to be planned and managed to achieve a sustainable system.

Conservation involves actions to use these resources most efficiently, thus extending their life as long as possible. For example, by **recycling** aluminium, the same piece of material is reused in a series of products, reducing the amount of aluminium ore that must be freshly mined. Conservation also involves the **reuse** of resources. Plastic bags are an obvious example, but there are many others. In many developing countries there has long been a culture of reuse and recycling because people cannot afford to buy new replacements.

Recycling also involves the recovery of waste. New technology makes it possible to recover mineral content from the waste of earlier mining operations. However, the proportion of a material recycled is strongly related to the cost in proportion to the price of the original raw material. Recycling not only conserves valuable resources, it is also fundamental in the reduction of landfill.

Various other strategies can be used in the attempt to conserve resources.

- The agreement of **quotas** is an increasingly frequent resource management technique, illustrated by the case study of the EU's Common Fisheries Policy.
- Much further along the line is **rationing**. This is very much a last-resort management strategy.
- At various times the use of **subsidies** has been criticised by environmentalists. It has been argued that reducing or abandoning some subsidies would aid conservation.

- Product stewardship is an approach to environmental protection in which manufacturers, retailers and consumers are encouraged or required to assume responsibility for reducing a product's impact on the environment.
- Substitution is the use of common and thus less valuable resources in place of rare, more expensive resources. An example is the replacement of copper by aluminium in the manufacture of a variety of products.

Energy efficiency

Meeting future energy needs while avoiding serious environmental degradation will require increased emphasis on approaches which include:

- much greater investment in renewable energy
- conservation
- recycling
- carbon credits
- 'green' taxation.

Managing energy supply is often about balancing socio-economic and environmental needs. **Carbon credits** and **carbon trading** are an important part of the EU's environment and energy policies. Under the EU's emissions trading scheme, heavy industrial plants have to buy permits to emit greenhouse gases over the limit they are allowed (carbon credits) by government. However, this could be extended to other organisations such as banks and supermarkets. Many countries are looking increasingly at the concept of **community energy**. Much energy is lost in transmission if the source of supply is a long way away. Energy produced locally is much more efficient. This will invariably involve **microgeneration**.

Table 3.15 summarises some of the measures governments and individuals can undertake to reduce the demand for energy and thus move towards a more sustainable situation.

▼ **Table 3.15** Examples of energy conservation measures

Government	Individuals
• Improve public transport.	• Walk rather than drive for short local journeys.
• Set a high level of tax on petrol.	• Buy low fuel consumption/low emission cars.
• Set minimum fuel consumption requirements for vehicles.	• Reduce car usage by planning more 'multi-purpose' trips.
• Congestion charging to deter non-essential car use in city centres.	• Use public rather than private transport.
• Encourage business to monitor and reduce its energy usage.	• Car pooling.
	• Use low-energy light bulbs.
• Promote investment in renewable forms of energy.	• Install and improve home insulation.
• Pass laws to compel manufacturers to produce more efficient electrical products.	• Turn boiler and radiator settings down.
	• Wash clothes at lower temperatures.
	• Purchase energy efficient appliances.

Case study: Environmental problems in the Pearl River delta

- The Pearl River delta region in south-east China is the focal point of a massive wave of foreign investment into China.
- The region's manufacturing industries already employ 30 million people. Major industrial centres include Shunde, Shenzhen and Guangzhou.
- The three major environmental problems in the Pearl River delta are air pollution, water pollution and deforestation.
- In 2007 eight out of every ten rainfalls in Guangzhou were classified as acid rain. The high concentration of factories and power stations is the source of this problem along with the growing number of cars in the province.
- Two-thirds of Guangdong's 21 cities were affected by acid rain in 2007. Overall, 45 per cent of the province's rainfall in 2007 was classified as acid rain.
- Almost all the urban areas have overexploited their neighbouring uplands, causing a considerable reduction in vegetation cover. This has resulted in serious erosion.
- Half of the wastewater in Guangdong's urban areas is not treated before being dumped into rivers. Guangdong's government is working to reduce chemical oxygen demand (COD) and also to cut sulfur dioxide emissions.
- The Environmental Protection Bureau classifies the environmental situation as 'severe'.
- Among the measures used to tackle the problems are (a) higher sewage treatment charges, (b) stricter pollution regulations on factories and (c) tougher national regulations on vehicle emissions.

Exam-style questions

1 a Define pollution. [2]
  b What are the main greenhouse gases? [2]
  c Explain the enhanced greenhouse effect. [4]
2 a Describe **two** possible consequences of enhanced global warming. [3]
  b Explain the causes of land degradation. [4]

Answers on pages 134–135

4.1 Geographical skills

Key definitions

REVISED

Term	Definition
Northings	The regular horizontal lines you can see on an Ordnance Survey map.
Eastings	The regular vertical lines you can see on an Ordnance Survey map.
Contour line	A line that joins places of equal height.
Cross-section	A view of the landscape as it would appear if sliced open, or if you were to walk along it.

Scale  PAGE 266

REVISED

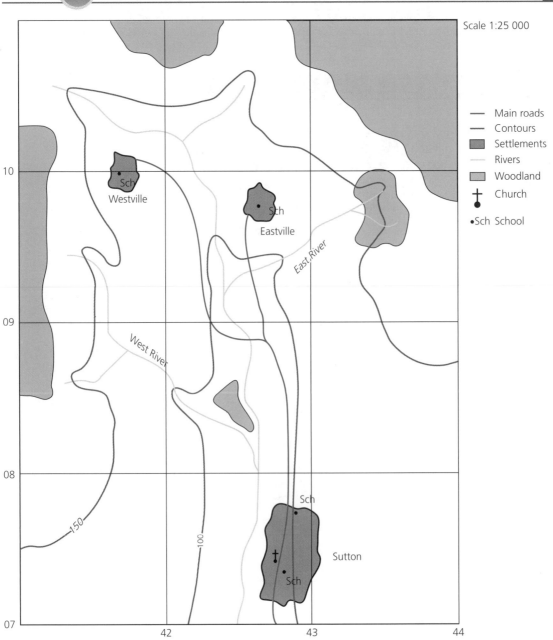

Scale 1:25 000

Key:
— Main roads
— Contours
▪ Settlements
— Rivers
▪ Woodland
✝ Church
•Sch School

▲ Figure 4.1

Most Ordnance Survey maps that are used are either at a 1:50 000 or a 1:25 000 scale. On a 1:50 000 map, 1 cm on the map relates to 50 000 cm on the ground. On a 1:25 000 map, every 1 cm on the map refers to 25 000 cm on the ground. In every kilometre there are 100 000 cm (1000 m × 100 cm). Hence:

● on a 1:50 000 map every 2 cm corresponds to a kilometre

● on a 1:25 000 map every 4 cm corresponds to a kilometre.

A 1:25 000 map is more detailed than a 1:50 000 map and is therefore an excellent source for geographical enquiries. 1:50 000 maps provide a more general overview of a larger area. You may come across other scales, for example, 1:10 000 and 1:2500.

Figure 4.1 shows a 1:25 000 map of the upper part of a river catchment. Notice that each of the boxes is exactly 4 cm × 4 cm. Figure 4.2 shows a 1:50 000 map of the same area (but only the contours are shown).

Measurement on maps is made easier by grid lines. These are the regular horizontal and vertical lines you can see on an Ordnance Survey map.

The horizontal lines are called **northings** and the vertical lines are called **eastings**. They help to pinpoint the exact location of features on a map.

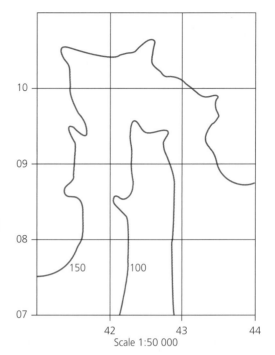

Scale 1:50 000

▲ **Figure 4.2**

Grid and square references PAGE 266

PAGE 266

REVISED

Grid references are the six-figure references which locate precise positions on a map. The first three figures are the eastings and these tell us how far a position is across the map. The last three figures are the northings and these tell us how far up the map a position is. An easy way to remember which way round the numbers go is 'along the corridor and up the stairs'.

In Figure 4.1, the confluence of the West River and East River (where they meet) is at 426080.

Sometimes a feature covers an area rather than a point, for example, all of the villages and the areas of woodland in Figure 4.1. Here a grid reference is inappropriate so we use four-figure square references.

● The first two numbers refer to the eastings.

● The last two numbers refer to the northings.

The point where the two grid lines meet is the bottom left-hand corner of the square. Thus in Figure 4.1, Eastville is located in 4209. Most of the village of Sutton is found in 4207. Some features may occur in two or more squares, for example, Westville is found in squares 4109 and 4110.

Test yourself

1 State the difference between *northings* and *eastings*.

2 State the difference between a *square reference* and a *grid reference*.

3 When would you use a *square reference* rather than a *grid reference*?

Answers on page 127

Direction PAGE 266

PAGE 266

REVISED

Directions can be expressed in two ways:

- compass points, for example, south-west
- compass bearings or angular directions, for example, 45°.

Sixteen compass points are commonly used. Some of these are shown in Figure 4.3.

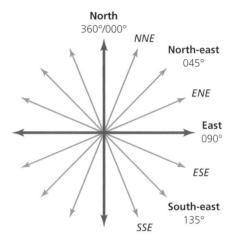

▲ **Figure 4.3** Compass points

Exam-style questions

Study Figure 4.1 on page 112.

1 State the grid reference for
   a the source of West River and
   b the school in Eastville.
2 State the square references for
   a Eastville and
   b the small area of woodland to the east of Eastville.
3 In which direction is
   a Eastville from Sutton and
   b Sutton from Westville?
4 What is the distance from the school at Eastville to
   a the school at Westville and
   b the school at 428074?
5 Complete Figure 4.2 by adding the woodland, rivers, main roads and settlements.

Answers on page 135

Relief and gradient PAGES 267–269

PAGES 267–269

REVISED

Contour lines

A **contour line** is a line that joins places of equal height.

- When the contour lines are spaced far apart the land is quite flat.
- When the contour lines are very close together the land is very steep (when the land is too steep for contour lines a symbol for a cliff is used).
- When contour lines are close together at the top, and then get further apart, it suggests a concave slope.
- When contour lines are close at the bottom and flat at the top, it suggests a convex slope.

Straight   Convex   Concave

▲ **Figure 4.4** Contour patterns

Gradients

The gradient of a slope is its steepness. We can get a rough idea of the gradient by looking at the contour pattern. We have just seen that if the contour lines are close together the slope is steep, and if they are far apart the land is quite flat. However, these are not very accurate descriptions. To measure gradient accurately we need two measurements:

- the vertical difference between two points (this can be worked out using the contour lines or spot heights)
- the horizontal distance between two places – this may or may not be a straight line (for example, a meandering stream would not be straight).

Test yourself

4 Explain what a *contour line* is.

5 Distinguish between a *convex* slope and a *concave* slope.

Answers on page 127

Exam-style questions

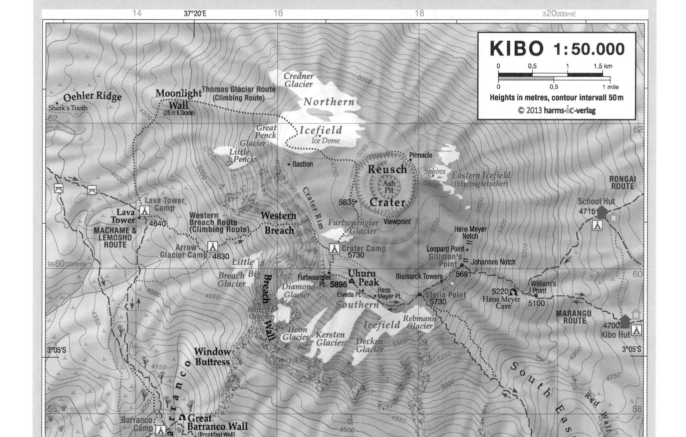

▲ **Figure 4.5** Mount Kibo 1:50 000 extract

Study Figure 4.5.

6 What is the approximate height of:
  a the Reusch Crater (1761) and
  b Uhuru Peak (170598)?

7 How far is it, in a straight line, from Uhuru Peak to Karanga Camp (171559)?

8 What is the approximate altitude of Karanga Camp?

9 Outline the map evidence to show that the gradient from Uhuru Peak to Karanga Camp is not a steady one.

Answers on page 135

Cross-sections PAGE 270

REVISED

A **cross-section** is a view of the landscape as it would appear if sliced open, or if you were to walk along it. It shows variations in gradient and the location of important physical and human features. To draw a cross-section:

1 Place the straight edge of a piece of paper between the two end points (Figure 4.6).

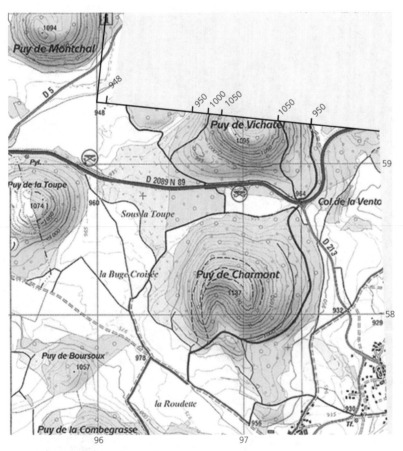

▲ **Figure 4.6** Drawing a cross-section

  a Mark off every contour line (in areas where the contours are very close together you could measure every second contour or significant contours, for example, every 100 m).

  b Mark off important geographical features.

2 Align the straight edge of the piece of paper against a horizontal line on graph paper, which is exactly the same length as the line of the section. Use a vertical scale of 1 cm:50 m or 1 cm:100 m; if you use a smaller scale (for example, 1 cm:5 m) you will end up with a slope that looks Himalayan!

a Mark off with a small dot each of the contours and the geographic features.

b Join up the dots with a freehand curve.

c Label the features.

d Remember to label the horizontal and vertical scales, the title and the grid references for the starting and finishing points.

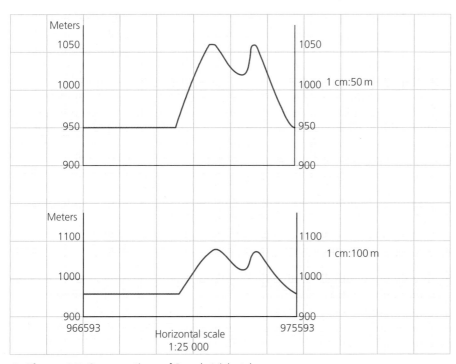

▲ **Figure 4.7** Cross-sections of Puy de Vichatel

Exam-style question

10 On Figure 4.8, draw a cross-section from 410090 to 440090 (from Figure 4.1) to show the rivers (drainage) and roads (transport). The base has already been drawn for you.

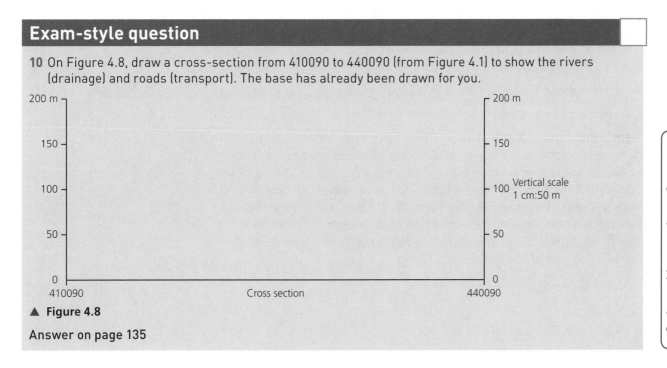

▲ **Figure 4.8**

Answer on page 135

mathematical skills in geography

 PAGES 281–284

Mean, averages and range

The **mean** is a type of average. It is found by totalling (Σ) the values for all observations (Σx) and then dividing by the total number of observations (n):

$$\Sigma \frac{x}{n}$$

For example, the number of services in eight villages was found to be

$$\frac{5+4+7+8+1+1+2+4}{8} = \frac{32}{8} = 4$$

The **range** is the difference between the highest and lowest value. In this example it is $8 - 1 = 7$.

Decimals, fractions, percentages and ratios

A **decimal** is a number between two whole numbers. For example, the world's population is around 7.5 billion, which means that it is more than 7 billion but less than 8 billion. A **fraction** is part of a whole – for example, $\frac{1}{3}$ of Borneo's rainforest has been deforested. A **percentage** is a number or ratio expressed as a fraction of 100 (%). To convert a fraction to a percentage, convert first to a decimal and multiply by 100. To convert a fraction to a decimal, simply divide the number above the line (the numerator) by the number below the line (the denominator). You can use a calculator for this. Thus, in Borneo $\frac{1}{3}$ or (0.33 × 100%) or 33% of its forest has been deforested. A **ratio** is a method of comparing relative sizes or proportions. The area deforested in Borneo compared with forested is 1:2 (one-third has been deforested and there are two-thirds left).

Standard notation, indices and significant figures

Standard notation is the number that we would normally write, for example, 567. The expanded standard index notation shows that 567 is 5.67×10^2.

A **positive** index is a power value that is positive, for example, $2^2 = 2 \times 2 = 4$, or $3^3 = 3 \times 3 \times 3 = 27$.

Negative **indices** are power values that have a minus sign, for example, $2^{-3} = \left(\frac{1}{2}\right)^3 = \frac{1}{8}$.

Significant figures are the numbers that carry some meaning to the measurement/size of a feature. Numbers are often rounded up or down to make them easier to understand. The world's population is said to be 7 billion – this is one significant figure, i.e. seven times a billion. The world's population was 7,503,875,592 at 20.30 hours, on 11 May 2017. This figure is too detailed (and out of date). A value of 7 billion (one significant figure) or 7.5 billion (two significant figures) gives a better 'feel' for the size of the world's population.

Test yourself

7 The world's population is approximately 7.5 billion and China's population is approximately 1.3 billion. Express China's population as a percentage of the world's population.

8 Express 7,503,875,592 in terms of:

   a four significant figures

   b seven significant figures

   c nine significant figures

Answers on page 127

Graphical skills PAGES 284–289

REVISED

Pie charts

Pie charts are subdivided circles. These are frequently used on maps to show variations in composition of a geographic feature, for example, the proportion of people living on less than $1.25/day in 1990 (Figure 4.9).

Plotting a pie chart

The following steps should be followed in the construction of a pie chart.

1 Convert the data into percentages.

2 Convert the percentages into degrees (by multiplying by 3.6 and rounding up or down to the nearest whole number).

3 Subdivide the circles into sectors using the figures obtained in step 2.

4 Differentiate the sectors by means of different shadings or colours.

5 Draw a key explaining the scheme of shading and/or colours.

6 Title the chart.

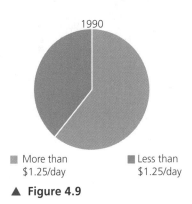

1990

■ More than $1.25/day   ■ Less than $1.25/day

▲ **Figure 4.9**

Exam-style question

11 Draw a pie chart to show the proportion of people living on less than $1.25/day in 2015.
   Percentage on less than $1.25/day = 14%
   Percentage on more than $1.25 = 86%

Answer on page 136

Bar charts

In a bar chart, the length of the bar represents the quantity of each component (for example, places or time intervals). The vertical axis has a scale that measures the total of each of these components. There are four main types of bar chart:

- **Simple bar charts** – each bar indicates a single factor. If the difference in length of bars is not great, then the difference can be emphasised by leaving a space between them or breaking the vertical scale.
- **Multiple or group bar chart** – features are grouped together on one graph to help comparison.

- **Compound bar chart** – various elements or factors are grouped together on one bar (the most stable element or factor is placed at the bottom of the bar to avoid confusion).
- **Percentage compound bar chart** – used to compare features by showing the percentage contribution. These graphs do not give a total in each category but compare relative changes in terms of percentages.

Exam-style question

Study Figure 4.10, which shows the populations of ten megacities in 2016.

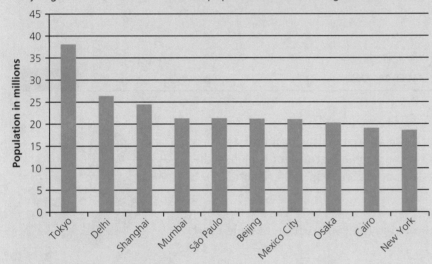

▲ **Figure 4.10** Populations of ten megacities, 2016 (millions)

12 Find the predicted population sizes for these ten megacities for 2030 from Table 1.21 (page 72) of the textbook, and illustrate these in a graph alongside their current size.

Answer on page 136

Scatter graphs PAGES 4–5

Scatter graphs show how two sets of data are related to each other, for example, population size and number of services, or distance from the source of a river and average pebble size. To plot a scatter graph decide which variable is independent (population size/distance from the source) and which is dependent (number of services/average pebble size). The independent is plotted on the horizontal or *x* axis (in Figure 4.11, GNI in $) and the dependent on the vertical or *y* axis (in Figure 4.11, the IMR).

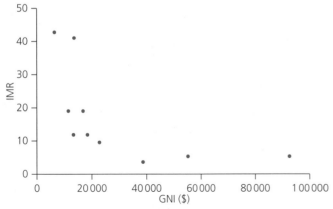

▲ **Figure 4.11** Scatter graph to show the relationship between GNI and IMR

A positive relationship exists when both variables increase, for example, as levels of atmospheric CO_2 increase, mean global temperatures increase (Figure 4.12a), whereas a negative relation exists when as one factor increases, the other decreases, for example, as gross national income increases, infant mortality decreases (Figure 4.12b).

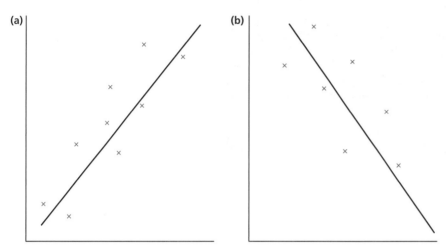

▲ **Figure 4.12** Relationships on a scatter graph: (a) positive relationship; (b) negative relationship

Exam-style question

13 Plot a scatter graph to show the relationship between GNI and life expectancy for the following countries:

Country	GNI ($)	Life expectancy (years)
UK	37 700	80.42
USA	54 800	79.56
China	12 900	75.15
India	5 800	67.80
Qatar	92 400	78.38
South Africa	12 700	49.50
Brazil	16 100	73.28
Egypt	11 100	73.45
Argentina	22 100	77.51
Mexico	17 900	75.43

Answer on page 136

Sketch maps and annotated photographs

 PAGE 280

REVISED

You can label a photograph or diagram to make it very informative. It is important that you label clearly all the important features.

The photograph (Figure 4.13) and sketch map (Figure 4.14) show part of a city's suburbs.

▲ **Figure 4.13** Aerial view of suburban development in Oxford

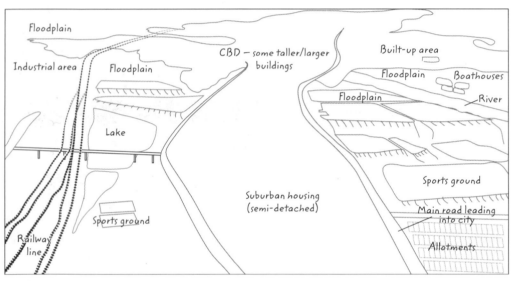

▲ **Figure 4.14** Sketch drawing of Figure 4.13

Exam-style question

14 Make a sketch of the photograph (Figure 4.15) to show the main features of the meandering stream.

▲ **Figure 4.15**

Answer on page 136

Sampling 📖 PAGES 294–295

The reasons for sampling

For many geographical investigations it is impossible to obtain 'complete' information. This is usually because it would just take too long in terms of both time and cost. For example, if you wanted to study the shopping habits of all the households in a suburban area by using a doorstep questionnaire, it would be a huge task to visit every household.

However, it is valid to take a 'sample' or proportion of this total 'population' of households, providing you follow certain rules. The idea is that you are selecting a group that will be representative of the total population.

Sampling types

Before selecting the sampling method you need to consider how you are going to take a sample at each location. There are three alternatives:

- Point sampling – making an observation or measurement at an exact location.
- Line sampling – taking measurements along a carefully chosen line or lines.
- Quadrat (or area) sampling – quadrats are mainly used for surveying vegetation and beach deposits.

Sampling methods

Random sampling

This method involves selecting sample points by using random numbers. Tables of random numbers can be used or the numbers can be generated by most calculators. The use of random numbers guarantees that there is no human bias in the selection process.

Systematic sampling

With this method the sample is taken in a regular way. It might, for example, involve every tenth house or person. When using an Ordnance Survey map it might mean analysing grid squares at regular intervals.

Stratified sampling

Here the area under study divides into different natural areas. For example, rock type A may make up 60 per cent of an area and rock type B the remaining 40 per cent. If you were taking soil samples for each type, you should ensure that 60 per cent of the samples were taken on rock type A and 40 per cent on rock type B.

Questionnaires PAGE 296

Questionnaire surveys involve both setting questions and obtaining answers. The questions are preplanned and set out on a specially prepared form. This method of data collection is used to obtain opinions, ideas and information from people in general or from different groups of people. The questionnaire survey is probably the most widely used method to obtain primary data in human geography. In the wider world, questionnaires are used for a variety of purposes, including market research by manufacturing and retail companies and to test public opinion prior to political elections.

One of the most important decisions you are going to have to make is how many questionnaires you are going to complete. The general rules to follow here are similar to those for sampling, set out in the previous section. Remember, if you have too few questionnaire results, you will not be able to draw valid conclusions. For most types of study, 25 questionnaires is probably the minimum you would need to draw reasonable conclusions. On the other hand it is unlikely you would have time for more than 100 unless you were collecting data as part of a group.

A good questionnaire:

- has a limited number of questions that take no more than a few minutes to answer
- is clearly set out so that the questioner can move quickly from one question to the next – people do not like to be kept waiting; the careful use of tick boxes can help this objective
- is carefully worded so that the respondents are clear about the meaning of each question
- follows a logical sequence so that respondents can see 'where the questionnaire is going' – if a questionnaire is too complicated and long-winded, people may decide to stop halfway through
- avoids questions that are too personal
- begins with the quickest questions to answer and leaves the longer/more difficult questions to the end
- reminds the questioner to thank respondents for their cooperation.

The disadvantages of questionnaires are:

- Many people will not want to cooperate for a variety of reasons. Some people will simply be too busy, others may be uneasy about talking to strangers, while some people may be concerned about the possibility of identity theft.
- Research has indicated that people do not always provide accurate answers in surveys. Some people are tempted to give the answer that they think the questioner wants to hear or the answer they think shows them in the best light.

Answers to test yourself questions

Topic 1.1

1. The difference between the birth rate and the death rate.
2. Africa
3. 2023
4. Demographic, social/cultural, economic, political
5. The average number of years a newborn infant can expect to live under current mortality levels.
6. Intense competition for land; heavy traffic congestion

Topic 1.2

1. A person forced to flee their home due to human or environmental factors, and who crosses an international border into another country.
2. In the nineteenth century and the early part of the twentieth century.
3. The process of population decentralisation as people move from large urban areas to smaller urban settlements and rural areas.

Topic 1.3

1. Stage 3, stage 5, stage 2, stage 4
2. The ratio of the number of people under 15 and over 64 years to those 15–64 years of age (per 100).
3. For every 100 people in the economically active population there are 80 people dependent on them.

Topic 1.4

1. The average number of people per square kilometre in a country or region.
2. Less than one person per km²
3. For example, Boston, New York, Chicago, Philadelphia

Topic 1.5

1. A dispersed settlement pattern occurs when farms or houses are set among fields or spread along roads. A nucleated settlement is one in which buildings are tightly clustered around a central feature.
2. The site of a settlement is the land on which the settlement is built, whereas the situation or position is the relationship between the settlement and its surrounding area.
3. High-order function – department store or bank; low-order function – newsagent
4. The threshold population is the number of people needed to support a good or service. The sphere of influence refers to the area that a settlement serves.

Topic 1.6

1. Urban sprawl is the unchecked outward spread of built-up areas, caused by their expansion. Urban renewal refers to the improvement of existing buildings.
2. Gentrification refers to the movement of higher social or economic groups into an area after it has been renovated and restored.

Topic 1.7

1. Urbanisation is the process by which the proportion of a population living in urban areas increases through migration and natural increase.
2. A megacity is a city with over of 10 million inhabitants, whereas a millionaire city is a city with over 1 million inhabitants.

Topic 2.1

1. A dormant volcano is a volcano that has not erupted for a very long time but could erupt again. In contrast, an extinct volcano is a volcano that has shown no signs of volcanic activity in historical times.
2. A destructive plate boundary is one where oceanic crust moves towards the continental crust and sinks beneath it, due to its higher density. Material is destroyed at a destructive plate boundary. In contrast, at a constructive plate boundary, new oceanic crust is formed as the two plates move apart.
3. For example, due to the weight of large dams, drilling for oil/fracking and/or nuclear testing.

Topic 2.2

1. Hydraulic action is the force of air and water on the sides of rivers and in cracks, whereas abrasion is the wearing away of the bed and bank by the load carried by a river.
2. Saltation is the process whereby heavier particles are bounced or bumped along the bed of the river, whereas traction is the process whereby the heaviest material is dragged or rolled along the bed of the river.
3. The main causes of floods are heavy rain, prolonged rain, snow melt, high tides, storm surges, earthquakes and landslides causing temporary lakes. Human activities can increase the flood risk by removing vegetation, making the surface impermeable and living in areas at risk of flooding.
4. The risk of flooding can be reduced by: building dams or reservoirs to hold back excess water; raising the banks of rivers; dredging the river channel so that it can hold more water; diverting streams and creating new flood relief channels; using sand bags to prevent water getting into houses; building houses on stilts so that water can pass underneath;

planning, i.e. build only on land that is ~~...~~ from flooding; afforestation (planting forests) to increase interception and reduce overland flow; having insurance cover for vulnerable areas and communities; improving forecasting and warning systems.

Topic 2.3

1 Destructive waves are erosional waves with a short wavelength and high height. They have a high frequency (10–12 per minute), and their backwash is greater than their swash. In contrast, constructive waves have a long wavelength and low height. They have a low frequency (6–8 per minute) and their swash is greater than backwash.

2 Steep cliffs are eroded between the low-water mark and the high-water mark. As they retreat they are replaced by a lengthening platform and lower-angle cliffs, subject to weathering and mass movements rather than marine forces.

3 Fringing reefs are those that grow outwards around an island, and are located on the edge of a land mass, whilst atolls are circular reefs enclosing a shallow lagoon.

4 Gabions are rocks held in wire cages and absorb wave energy, whereas sea walls are large-scale concrete structures designed to deflect wave energy.

Topic 2.4

1 A Stevenson screen is a wooden box standing on four legs of height about 120 cm. The screen is raised so that air temperature can be measured. The sides of the box are slatted to allow air to enter freely. The roof is usually made of double boarding to prevent the Sun's heat from reaching the inside of the screen. Insulation is further improved by painting the outside of the screen white to reflect much of the Sun's energy. The screen is usually placed on a grass-covered surface, thereby reducing the radiation of heat from the ground.

2 a Wet- and dry-bulb thermometer

   b Anemometer

3 Cirrus clouds are thin, wispy, high-altitude clouds, formed of ice crystals. Cumulonimbus are thick clouds extending from low altitude to high altitude, and are associated with heavy downpours and sometimes thunder and lightning.

Topic 2.5

1 a Equatorial areas have: high annual temperatures year round (26–27°C); low seasonal ranges (1–2°C), but greater diurnal (daily) ranges (10–15°C); high rainfall throughout the year (more than 2000 mm).

   b Hot deserts have: high daytime temperatures (30–40°C) throughout the year; a large temperature difference, often as much as 50°C between day and night; unreliable and low rainfall (about 250 mm year).

2 a Areas that are close to the Equator receive more heat than areas that are close to the Poles

because incoming solar radiation is concentrated near the Equator, but dispersed near the Poles. In addition, insolation near the Poles has to pass through a greater amount of atmosphere and there is more chance of it being reflected back out to space.

   b Temperature decreases with altitude. On average, it drops about 1°C for every 100 metres.

3 Vegetation in the rainforest is evergreen. Photosynthesis and growing can happen all year. The vegetation is layered, and the shape of the crowns varies with the layer, in order to receive maximum light. It is very diverse — there are up to 200 species of tree per hectare. Trees at the top of the canopy are adapted to being in the light; those near the ground are adapted to being in the shade. The trees have buttress roots to help stabilise them.

4 Plants adapt to deserts by: growing deep roots or wide roots; producing only a few leaves, to reduce transpiration (moisture loss); storing water; and producing seeds that wait for rainfall before growing and completing a very quick life cycle.

Topic 3.1

1 Life expectancy at birth, mean years of schooling for adults aged 25 years, expected years of schooling for children of school entering age, GNI per capita (PPP$)

2 48

3 A technique used to show the extent of income inequality.

4 The full sequence of activities needed to turn raw materials into a finished product.

5 A firm that owns or controls productive operations in more than one country through foreign direct investment.

6 From 361 million in 2000 to 3.4 billion in 2016

Topic 3.2

1 Inputs, processes, outputs

2 Livestock farming such as dairy cattle, beef cattle, sheep and pigs

3 The way in which land is, or can be, owned.

Topic 3.3

1 Industries that are not tied to certain areas because of energy requirements or other factors.

2 For example, Cambridge Science Park in the UK

3 For example, capital, labour, markets

Topic 3.4

1 2012

2 A particular location where economic development, such as tourism, is focused, setting off wider growth in the surrounding region.

3 The part of the money a tourist pays for a foreign holiday that does not benefit the destination country because it goes elsewhere.

Topic 3.5

1 Sources of energy that are not depleted as they are used.
2 About 2.5 billion people
3 China, the USA, Germany, India and Spain

Topic 3.6

1 Over 660 million
2 Desalination
3 When water supply falls below 1000 cubic metres per person a year, a country faces water scarcity for all or part of the year.

Topic 3.7

1 See Figure 3.20 on page 104.
2 $450\,km^3$ globally
3 Preventing its occurrence; repairing the damage
4 For example, greater global temperature variations, rising sea levels, melting of ice caps and glaciers
5 Deforestation and overgrazing
6 The control of the exploitation and use of resources in relation to environmental and economic costs.

Topic 4.1

1 Northings are the horizontal lines on a map. They show how far north (or south) a place is. Eastings are the vertical lines on a map. They show how far east (or west) a place is.
2 Square references are four-figure references, whereas grid references are six-figure references.
3 Square references are used to show areas, whereas grid references are used to show precise points.
4 A contour line is a line that joins places of equal height.
5 When contour lines are close together at the top, and then get further apart nearer the bottom, it is a concave slope. In contrast, when contour lines are close at the bottom and flatter near the top, it is a convex slope.
6 A view of the landscape as it would appear if sliced open, or a side-on view of the path followed if you were to walk along it. It shows variations in gradient and altitude.
7 17%
8 a 7,504,000,000
  b 7,503,876,000
  c 7,503,875,590

Answers to exam-style questions

Topic 1.1

1 a The birth rate is the number of live births per 1000 population per year in a country or region. The total fertility rate is the average number of children women have during their lifetimes in a country or region.

   A clear and accurate distinction between the two definitions.

 b Africa has the highest total fertility rate, while Europe has the lowest.

 c Unlike the crude birth rate the total fertility rate takes account of both age structure and gender. It focuses only on women in the reproductive age range. It is thus a much more accurate measure of fertility than the birth rate.

   Clearly identifies the key elements in the total fertility rate (TFR).

2 a Where infant mortality is high, it is usual for many children to die before reaching adult life. In such societies, parents often have many children to compensate for these expected deaths. In many poor countries children are viewed as an economic asset. The infant mortality rate for the world as a whole was 41/1000 in 2012, ranging from 5/1000 in Europe to 67/1000 in Africa. It is not just coincidence that the continent with the lowest fertility is Europe and the continent with the highest fertility is Africa. There is a strong relationship between the infant mortality rate and the total fertility rate.

   An answer with good sequence and progression, and good use of relevant data.

 b Education, especially female literacy, is seen by most experts as the key to lower fertility. With education comes a knowledge of birth control, greater social awareness, more opportunity for employment and a wider choice of action generally for women. When women have wider choices they generally marry later, start their families later and also have fewer children. Figure 1.9 shows that there is a strong negative correlation on the scatter graph which compares a large number of countries. As the percentage of girls in secondary education increases, the total fertility rate falls.

   Shows clear understanding of the relationship between the two variables with accurate reference to the trend shown on the scatter graph.

Topic 1.2

1 a Migration is the movement of people across a specified boundary, national or international, to establish a new permanent place of residence.

 b In voluntary migration, the individual has a free choice about whether to migrate or not. In involuntary migrations, people are made to move against their will and this may be due to human or environmental factors. An example of a human factor would be 'ethnic cleansing'. An example of an environmental factor would be a volcanic eruption.

   A good, clear distinction with relevant examples.

 c Prior to 1914, government controls on international migration were almost non-existent. The main obstacles to migration at the time were financial cost and the physical dangers associated with the journey, for example, the long voyage across the Atlantic Ocean from Europe to America in the eighteenth and nineteenth centuries. Since the early twentieth century, government controls on migration have been steadily tightened. Thus, immigration restrictions are by far the greatest barrier to legal migration today. Most governments are looking for migrants whose skills they need and who are likely to make a strong contribution to their new country. However, other barriers to migration remain, such as language, lack of awareness of opportunities and intolerance in receiving countries. For illegal migrants, the financial costs and physical dangers can be substantial as they try to avoid the barriers to entry. Thousands of people have died on the journey across the Mediterranean Sea from North Africa to Europe when overcrowded boats run by people traffickers have sunk.

   Very good understanding of the way in which the barriers to migration have changed over time with excellent reference to recent illegal migration and the difficulties involved.

2 a Push factors are negative conditions at the point of origin which encourage or force people to move. For example, a high level of unemployment is a major push factor in a region or a country. In contrast, pull factors are positive conditions at the point of destination which encourage people to migrate. An important pull factor is often much higher wages in another country or region. The nature of push and pull factors varies from country to country (and from person to person) and changes over time.

   Clear distinction between push and pull factors with relevant examples.

 b One of the largest labour migrations in the world has been from Mexico to the USA. This migration has largely been the result of:

 • much higher average incomes in the USA

 • lower unemployment rates in the USA

 • the faster growth of the labour force in Mexico, with significantly higher population growth in Mexico compared with the USA

 • the overall difference in the quality of life: on virtually every aspect of the quality of life, conditions are better in the USA than Mexico.

All of these factors have made the USA an attractive destination for migrants from Mexico. Over time, large Mexican communities have developed, particularly in the four states along the US/Mexican border: California, Arizona, New Mexico and Texas. These communities provide important networks for new would-be migrants. Most migration between Mexico and the USA has taken place in the last three decades. Although previous surges occurred in the 1920s and 1950s when the US government allowed the recruitment of Mexican workers as guest workers, persistent mass migration between the two countries did not take hold until the late twentieth century.

A detailed answer identifying the relevant factors with good case study information.

Topic 1.3

1 a The two aspects of population structure shown in a population pyramid are age and gender.

  b The first line should be drawn between the 10–14 and 15–19 age bars to show the division between the young dependent population and the economically active population. The second line should be drawn between the 60–64 and 65–69 bars to show the divide between the economically active population and the elderly dependent population.

  Accurately identifies the dividing lines between the three sections of a population.

  c The very wide base of the Gambia's population pyramid illustrates its very young population: 44 per cent of the population are classed as young dependents as they are under 15 years of age. The considerable reduction in the width of each successive bar with increasing age indicates high mortality and relatively low life expectancy. There are very few people indeed in the older age groups. Only 2 per cent of the population are over 65 and classed as elderly dependents. This gives a dependency ratio of 85. This means that for every 100 people in the economically active population in the Gambia there are 85 people dependent on them.

  A clear sequence of discussion with detailed analysis of the population pyramid.

2 a The dependency ratio is the relationship between the working or economically active population and the non-working population.

  b For every 100 people in the economically active population there are 80 people dependent on them.

  c The dependency ratio in developed countries is usually between 50 and 75, with the elderly forming an increasingly high proportion of dependents. Developing countries typically have higher dependency ratios, which can reach over 100, with young people making up the majority of dependents.

  Good understanding, with reference to data.

Topic 1.4

1 a Population density is the average number of people per square kilometre in a country or region. Population distribution is the way that the population is spread out over a given area, from a small region to the Earth as a whole.

  b About 80 per cent of Canada has a population density of less than one person per square kilometre. Areas with a similar low density within the USA are confined to the mountainous western part of the country. The highest population density shown on the map, over 100 per km^2, is confined to one coastal and one inland zone in the northeast of the country. Most areas with a density between 20 and 100 are in the eastern part of the USA. The exceptions are in Canada between Quebec and Toronto, on the west coast of the USA, and in a few other isolated areas.

  A good structured analysis with clear reference to the key.

2 a People have always tried to avoid harsh environments where making a living is particularly difficult. Thus, the world's lowest population densities are found in deserts (the Sahara desert), high mountain regions (the Himalayas), very cold landscapes (Canadian Northlands) and rainforests (the Amazon basin). In areas which are, in general, more hospitable, poor soil fertility can deflect people to areas of more fertile soils. Likewise, restricted water supply can deter settlement.

  Good identification of the factors resulting in low population density and reference to relevant examples.

  b In the USA the greatest concentration of population is in the north-east, the first region of substantial European settlement. By the end of the nineteenth century it had become the greatest manufacturing region in the world. The region stretches inland from Boston and Washington to Chicago and St Louis. By the 1960s the very highly urbanised area between Boston and Washington had reached the level of a megalopolis. This is the term used to describe an area where many conurbations exist in relatively close proximity. The region is sometimes referred to as 'Boswash' after the main cities at its northern and southern extremities. Apart from Boston and Washington, the other main cities in this region are New York, Philadelphia and Baltimore.

  New York is classed as a 'global city', being one of the world's three great financial cities along with Tokyo and London. With a population of 8.4 million, New York is the most densely populated major city in the USA. The population of the larger Metropolitan Area of New York is 18.9 million.

  The region also contains many smaller urban areas. Much of the area has an average density over 100 per km^2. Population densities are, of course, much higher in the main urban areas. The rural parts of the region are generally fertile and intensively farmed. The climate and soils at this latitude are conducive to agriculture. Many people living in the rural communities commute to work in the towns and cities. The region has the most highly developed transport networks in North America. Although other parts of the country are growing at a faster rate, the intense concentration of job opportunities in the north-east will ensure that it remains the most densely populated part of the continent in the foreseeable future.

A detailed answer showing good, clear case study knowledge. Relevant data included.

Topic 1.5

1 A linear settlement is spread along a single road whereas a nucleated shape is one which is compact and concentrated around a number of roads.

Sound definitions.

2 Generally, as population size increases, the number of services increases. This is a positive relationship. However, there are two main anomalies. Dormitory settlements are settlements with a relatively large population but limited services. In contrast, tourist locations may have a small population but a large number of services, for example, Villefort and St-André-Capcèze.

Accurate description and useful use of data.

Topic 1.6

1 Air pollution is likely, as there is very little open space so wind/breezes will be limited. There is likely to be much traffic congestion – there is a high density of buildings, and many tall buildings, and the space for roads appears to be limited. There is likely to be a high level of noise due to the combination of vehicles and large population density.

Logical suggestions.

2 a It increased from about 0.5 million in 1900 to a peak of about 1.8 million in 1950, and then declined to around 0.7 million in 2013.

Good use of data.

  b The decline in the car industry has left Detroit far less attractive for migrants, as there are fewer jobs available. As there are fewer jobs, many residents have left the area, too. Also, the 'white flight' of the 1950s to 1970s led to a fall in the population.

Identifies the main reasons.

Topic 1.7

1 Rapid urbanisation is caused by economic development and the perceived availability of employment and a better standard of living in urban areas, combined with poorer opportunities in rural areas. In addition, as a result of rural–urban migration, the urban area has a younger population structure than the rural area, and so has a higher rate of natural increase.

Covers both aspects clearly.

2 For people living in a shanty town (for example, Vidigal in Rio de Janeiro), they have a house or an apartment. They may also be a part of a community

that looks after each other, especially with respect to education. They may also be able to tap into electricity from the electricity power supply. However, there are many negative aspects of living in a squatter settlement. The quality of housing is poor, the nature of the jobs available is limited and the pay is quite poor. Access to healthcare and formal education is limited. Nevertheless, there are differences between squatter settlements. Those that are closer to the city centre offer more jobs. These are known as 'slums of hope'. In contrast, squatter settlements on the edge of the city may be far from centres of employment, and may be known as 'slums of despair'.

Covers both aspects clearly – an introductory sentence stating 'The advantages of living in a squatter settlement include …' would be useful.

3 The government could legalise the squatter settlements; they could provide a site and service scheme to help people without homes; they could improve the infrastructure such as running water and sanitation; they could provide loans for residents to improve their housing.

Identifies a range of possibilities.

Topic 2.1

1 Cone volcanoes are steep sided and formed of acidic ash and cinders, and are formed at destructive plate boundaries. In contrast, shield volcanoes are low-angled volcanoes – they may still be very high – formed of runny, basaltic lava at constructive plate margins and hot spots.

A clear answer.

2 The focus is the exact position within the Earth where an earthquake takes place. The epicentre is the point on the Earth's surface immediately above the focus.

Two clear definitions.

3 The advantages of volcanoes include: the creation of new land; the production of fertile soil; rich minerals; and they are important tourist destinations.

Four advantages identified.

4 The primary hazards are the direct hazards associated with earthquakes such as land shaking and surface faulting (subsidence). In contrast, secondary hazards are the indirect hazards such as landslides, tsunamis and rock falls.

Two definitions, both with support.

Topic 2.2

1 Infiltration is the movement of water into the soil, whereas throughflow is the downward movement of water under the soil (subsoil).

Clear definitions.

2 An upper course river has a narrow, deep/steep-sided cross-section, whereas a lower course river has a wider, flatter cross-section.

Clear description.

3 The upper course generally has a steeper long profile, whereas the lower course has a much gentler long profile.

Clear description.

Topic 2.3

1 Longshore drift occurs when the waves break at an angle to the shoreline. The swash moves up the beach at an angle whereas the backwash moves sediment down the beach at right angles to the shoreline. The net movement is along the beach, i.e. longshore drift.

A clear description.

2 On a rocky headland, hydraulic action erodes a line of weakness (such as a fault line) or wave refraction concentrates wave energy on the flanks of the headland, causing a cave to be formed. Continual erosion may, in time, form an arch in the headland. Over time the arch is weathered and eroded, until eventually it collapses, leaving a stack.

A good description, with the use of geographical terminology.

3 A spit is formed by longshore drift. It carries material and deposits it where there is an indent in the coastline or a river mouth causes an obstruction to longshore drift. A spit is attached at one end and unattached at the other end.

A good answer, although a diagram would help to show why longshore drift occurs.

4 Hazards include a combination of high wind speeds, storm surges (high tides), heavy rainfall leading to flooding and wind damage.

An appropriate answer.

5 The Nile Delta is low lying and a great many people live there. These factors, as well as it being an important agricultural and industrial region, make it vulnerable to sea level rising.

Clear reasons for vulnerability.

6 It is low lying, therefore easy to build on; it is fertile, so good for farming; it has a good supply of fresh water; it has good potential for trade because the river is navigable and has access to the coast; and it has good potential for tourism.

A range of reasons given.

Topic 2.4

1 Manaus: 27 °C (accept 26 °C–28 °C)
  Cairo: 33 °C (accept 32 °C–34 °C)

2 Manaus: 27 °C (accept 26 °C–28 °C)
  Cairo: 15 °C (accept 14 °C–16 °C)

3 Manaus: 1 °C (accept 1 °C–2 °C)
  Cairo: 21 °C (accept 20 °C–22 °C)

4 Manaus has maximum rainfall in summer (it is in the southern hemisphere), and has a high total annual rainfall. It has 200 mm of rain or more in each of the months from December to April. Cairo has negligible rainfall throughout the year – a mere 25 mm per annum.

Topic 2.5

1 Tropical rainforests are hot because they are located close to the Equator (low latitude), and therefore receive high levels of insolation throughout the year. They are wet because the high temperatures lead to convectional rainfall throughout the year. As you move away from the Equator, the climate becomes more seasonal.

A good answer – a few figures, for example, 27 °C year-round and rainfall of over 2000 mm, would make this answer much stronger.

2 Most of the nutrients are locked in the vegetation due to the year-round growing season. As leaves fall from the trees (there is no season), they are rapidly decomposed in the hot, wet conditions (over 27 °C and over 2000 mm per year). Due to the continuous growth of the vegetation, plants take up the nutrients, thereby leaving the soil relatively infertile.

3 The lack of water is the main limiting factor. Annual rainfall is less than 250 mm of rain, and aquifers in desert areas are non-renewable resources. Desalination offers a potential solution for the storage of water, but is expensive.

Topic 3.1

1 a The primary sector exploits raw materials from land, water and air. Farming, fishing, forestry, mining and quarrying make up most of the jobs in this sector.

  b The poorest countries of the world have more than 70 per cent of their employment in the primary sector. Lack of investment in general means that agriculture and other parts of the primary sector are very labour intensive, and jobs in the secondary and tertiary sectors are limited in number.

  c In NICs employment in manufacturing has increased rapidly in recent decades. NICs have reached the stage of development whereby they attract foreign direct investment from TNCs in many sectors of the economy. As NICs expand their economies they develop their own domestic companies. Such companies usually start in a small way, but some go on to reach a considerable size. Both processes create employment in manufacturing and services. The increasing wealth of NICs allows for greater investment in agriculture. This includes mechanisation which results in falling demand for labour on the land. So, as employment in the secondary and tertiary sectors rises, employment in the primary sector falls. NICs may become so advanced that the quaternary sector begins to develop.

  Shows detailed understanding of the ways in which employment structure has changed in NICs in recent decades.

2 a A transnational corporation is a firm that owns or controls productive operations in more than one country through foreign direct investment.

  b TNCs and nation states are the two main elements of the global economy. The governments of countries individually and collectively set the rules for the global economy, but the bulk of investment is through TNCs. Under this process, manufacturing industry at first, and more recently services, have relocated in significant numbers from developed countries to selected developing countries as TNCs have taken advantage of lower labour costs and other ways to reduce costs. It is this process which has resulted in the emergence of an increasing number of newly industrialised countries since the 1960s. It has resulted in deindustrialisation in many regions of the developed world.

Twenty years ago, the vast majority of the world's TNCs had their headquarters in North America, Western Europe and Japan. However, over the last two decades NICs such as South Korea, China and India have been accounting for an increasing slice of the global economy. Much of this economic growth has been achieved through the expansion of their own most important companies. TNCs have a huge impact on the global economy in general and in the countries in which they choose to locate in particular. They play a major role in world trade in terms of what and where they buy and sell. They are major employers and can have a huge influence on the countries in which they locate.

Clear understanding of the respective roles of TNCs and nation states, with important reference to the development of NICs and resultant deindustrialisation in developed countries.

Topic 3.2

1 a   Individual farms can be seen to operate as a system with inputs, processes and outputs. A farm requires a range of inputs such as labour so that the processes that take place on the farm, such as harvesting, can be carried out. The aim is to produce the best possible outputs such as milk, eggs and crops.

The elements of a system are listed, followed by an example of each element.

   b i   Intensive farming is characterised by high inputs per unit of land to achieve high yields per hectare. Extensive farming is where a relatively small amount of agricultural produce is obtained per hectare of land, so such farms tend to cover large areas of land. Inputs per unit of land are low.

     ii   Subsistence farming is the most basic form of agriculture where the produce is consumed entirely or mainly by the family who work the land or tend the livestock. If a small surplus is produced it may be sold or traded. The objective of commercial farming is to sell everything the farm produces. The aim is to maximise yields to achieve the greatest profit.

2 a   Precipitation is low in the winter months. Between November and April, each month has less than 50 mm. Precipitation increases considerably in May to about 130 mm, reaching over 300 mm in each of July and August (the peak month). After this peak, monthly precipitation falls rapidly.

The annual variation in temperature is much less than that for precipitation. December and January are the coldest months at about 19 °C and 20 °C respectively. Temperature then rises to 28 °C in April with little variation for the next five months, before falling steadily to the end of the year.

Detailed analysis of the climate graph in terms of both temperature and precipitation with clearly focused explanation.

   b   Temperature is a critical factor in crop growth as each type of crop requires a minimum growing temperature and a minimum growing season. Latitude, altitude and distance from the sea are the major influences on temperature. Precipitation is equally important. This is not just the annual total but the way it is distributed throughout the year. Long, steady periods of rainwater to infiltrate into the soil are best, making water available for crop growth. In contrast, short heavy downpours can result in surface runoff, leaving less water available for crop growth, and soil erosion.

Topic 3.3

1 a   Processing industries are based on the direct processing of raw materials. The iron and steel industry is an example, using large quantities of iron ore, coal and limestone. Processing industries are often located close to their raw materials. In contrast, assembly industries put together parts and components which have been made elsewhere. A large car assembly plant will use thousands of components to build a car. Assembly industries are more footloose in their choice of location.

   b   High-technology industry uses or makes silicon chips, computers, software, robots, aerospace components and other very technically advanced products. These companies put a great deal of money into scientific research. Their aim is to develop newer, even more advanced products.

   c   High-technology industries often cluster together in science parks which were originally created in the USA. They are often found in close proximity to leading universities because of the need to employ well-qualified graduates in science and technology, and to be aware of the latest research taking place in universities. The Cambridge Science Park is a major example in the UK. The clustering of high-technology industry means that companies can collaborate easily on joint projects and highly skilled workers can move easily from one company to another.

2   Bangalore is the most important city in India for high-technology industry. Bangalore's pleasant climate is a significant attraction to foreign and domestic companies alike. Because of its dust-free environment, large public sector undertakings, such as Hindustan Aeronautics Ltd and the Indian Space Research Organisation, were established in Bangalore by the Indian government. In addition, the state government has a long history of support for science and technology. There are many colleges of higher education in this sector and there has been large-scale investment in science and technology parks. The city prides itself on a 'culture of learning' which gives it an innovative leadership within India.

In the 1980s Bangalore became the location for the first large-scale foreign investment in high technology in India when Texas Instruments selected the city above a number of other possibilities. Other TNCs soon followed as the reputation of the city grew. Important backward and forward linkages were steadily established over time. Apart from ICT industries, Bangalore is also India's most important centre for aerospace and biotechnology.

India's ICT sector has benefited from the filter down of business from the developed world. Many European and North American companies which previously outsourced their ICT requirements to local companies are now using Indian companies. Outsourcing to India occurs because: labour costs are considerably lower; a number of developed countries have significant ICT skills shortages; India has a large and able English-speaking workforce.

A good answer with a clear sequence of discussion, showing detailed case study knowledge.

Topic 3.4

1 a Tourism is travel away from the home environment (i) for leisure, recreation and holidays, (ii) to visit friends and relatives, and (iii) for business and professional reasons.

A precise definition – students can sometimes be very vague when defining tourism.

b Steadily rising real incomes have enabled more and more people, particularly in developed countries, to afford a holiday away from home. An increasing number of people take more than one holiday trip each year. An increase in the average number of days of paid leave has given people more time in which to travel. And holiday pay helps with cost of travel. People have raised expectations of international travel with increasing media coverage of holidays, travel and nature. The range of available holidays has increased so that virtually all income levels are catered for.

c The traditional cultures of many communities in the developing world have suffered because of the development of tourism. The disadvantages include the following: the loss of locally owned land; the abandonment of traditional values; displacement of people; the weakening of traditional community structures; the increasing availability of alcohol and drugs; crime and prostitution, sometimes involving children; visitor congestion at key locations; denying local people access to beaches; the loss of housing for local people as more visitors buy second homes.

However, tourism may also generate cultural advantages such as: leading to greater understanding between people of different cultures; increasing the range of social facilities for local people; helping to develop foreign language skills in host communities.

A well-organised answer that does not just concentrate on the disadvantages, but also recognises that there can be some social/cultural advantages.

2 a The carrying capacity is the number of tourists a destination can take without placing too much pressure on local resources and infrastructure. If the carrying capacity of a tourist location is soon to be reached, important decisions have to be made. Will measures be taken to restrict the number of tourists to remain within the carrying capacity or are there possible management techniques that will allow the carrying capacity to be increased, but continue to be sustainable?

b Ecotourism is a specialised form of tourism where people experience relatively untouched natural environments such as coral reefs, tropical forests and remote mountain areas, and ensure that their presence does no further damage to these environments. In Ecuador, ecotourism has helped to bring needed income to some of the poorest parts of the country. It has provided local people with a new, alternative way of making a living. As such it has reduced human pressure on ecologically sensitive areas. The main geographical focus of ecotourism has been in the Amazon rainforest around Tena, which has become the main access point. The ecotourism schemes in the region are usually run by small groups of indigenous Quichua Indians.

A detailed and accurate definition followed by a succinct reference to a relevant case study.

Topic 3.5

1 a The energy mix of a country is the relative contribution of different energy sources to a country's energy consumption.

b The highest consumption countries, such as the USA, Canada and Saudi Arabia, use more than 6 tonnes oil equivalent per person, while almost all of Africa and much of South America and Asia use less than 1.5 tonnes oil equivalent per person. Other high energy-use regions, with consumption between 4.5 and 6.0 tonnes of oil equivalent, include Russia, Australia and Scandinavia.

There is a strong relationship between energy usage and the wealth of individual countries. Richer countries can afford to use more energy while poorer nations are greatly restricted by the cost of energy. However, climate is also an important influence, which helps to explain the high consumption of energy in Russia.

A good balance between description and explanation, with clear reference to the map key.

c In developing countries about 2.5 billion people rely on fuelwood, charcoal and animal dung for cooking. Fuelwood and charcoal are collectively called fuelwood, which accounts for just over half of global wood production. Fuelwood provides much of the energy needs for sub-Saharan Africa. It is also the most important use of wood in Asia.

2 No other source of energy creates such heated discussion as nuclear power. A major concern about nuclear power is the risk of power plant accidents, which could release radiation into air, land and sea. Radioactive waste storage/disposal is another big

problem. Most concern is over the small proportion of 'high-level waste'. This is so radioactive it generates heat and corrodes all containers. It would cause death within a few days to anyone directly exposed to it. No country has yet implemented a long-term solution to the nuclear waste problem.

There are serious concerns about rogue state or terrorist use of nuclear fuel for weapons. As the number of countries with access to nuclear technology rises, such concerns are likely to increase. High construction and decommissioning costs mean that the investment required is very high with only a limited number of countries being able to afford it. Because of the genuine risks associated with nuclear power and the level of security secrecy required, it is seen by some people as less 'democratic' than other sources of power.

There has also been debate about the possible increase in certain types of cancer near nuclear plants. There has been much debate about this issue, but the evidence appears to be becoming more convincing.

A good sequence of discussion covering all the major concerns about nuclear power.

Topic 3.6

1   a   In about 80 countries, with 40 per cent of the world's population, lack of water is a constant threat. And the situation is getting worse, with demand for water doubling every 20 years. In those parts of the world where there is enough water, it is being wasted, mismanaged and polluted on a large scale. In the poorest nations it is not just a question of lack of water; the paltry supplies available are often heavily polluted.

    b   Water supply is the provision of water by public utilities, commercial organisations or by community endeavours. The objective in all cases is to supply water from its source to the point of usage.

    c   In the twentieth century, global water consumption grew sixfold, twice the rate of population growth. Much of this increased consumption was made possible by significant investment in water infrastructure, particularly dams and reservoirs, affecting nearly 60 per cent of the world's major river basins. The world's major dams are really massive structures capable of holding huge amounts of water in the reservoirs behind them. This water can be released gradually as and when required by the settlements downstream of the dam. Reservoir storage needs have increased as world population has grown. There are approximately 80 000 dams of varying sizes in the USA alone. Globally, the construction of dams has declined since the height of the era in the 1960s and 1970s. This is because most of the best sites for dams are already in use or such sites are strongly protected by environmental legislation and, therefore, off-limits for construction. An alternative to building new dams and reservoirs is to increase the capacity of existing reservoirs by extending the height of the dam.

A good understanding of the importance of dams and reservoirs to global water supply, backed up by some useful data.

2   a   A well or borehole is a means of tapping into various types of aquifers (water-bearing rocks), gaining access to groundwater. Thus, they are sunk directly down to the water table. The water table is the highest level of underground water. For many communities, groundwater is the only water supply source. Aquifers provide approximately half of the world's drinking water, 40 per cent of the water used by industry and up to 30 per cent of irrigation water.

    Typically, a borehole is drilled by machine and is relatively small in diameter. Wells are relatively large in diameter and are often sunk by hand, although machinery may be used. Water from groundwater sources can be used directly or stored to build up a considerable surface supply. The greatest reliance on groundwater is in arid and semi-arid areas. This is the main source of water of oasis settlements such as those in the Sahara Desert in North Africa.

    b   Forest water management can be very important in many areas. Land management activities can affect water flow and degrade the quality of waters. Many countries rely on 'protection forests' to preserve the quality of drinking water supplies, alleviate flooding and to guard against erosion, landslides and the loss of soil.

    Households may be encouraged to use water butts to trap rainwater by other methods, thus taking less from the piped public supply. They may also be encouraged to use 'grey water' to water gardens. Grey water is water that has already been used, such as bath water.

Topic 3.7

1   a   Pollution is contamination of the environment. It can take many forms – air, water, soil, noise, visual and others.

    b   The main greenhouse gases being created by human activity are carbon dioxide, methane, nitrous oxides, chlorofluorocarbons and ozone.

    c   The Earth–atmosphere system has a natural greenhouse effect that is essential to all life on Earth, but large-scale pollution of the atmosphere by economic activities has created an enhanced greenhouse effect. The increase in greenhouse gases due to human activity is causing more radiation from the Earth's surface to be trapped in the atmosphere. This is causing temperatures to increase beyond the limits of the natural greenhouse effect. Many parts of the world are experiencing changes in their weather that are unexpected. Some of these changes could have disastrous consequences for the populations of the areas affected, if they continue to get more severe.

A good clear answer, following a logical sequence of argument.

2 a A major consequence of enhanced global warming is the melting of ice caps and glaciers. Satellite photographs show ice melting at its fastest rate ever. The area of sea ice in the Arctic Ocean has decreased by 15 per cent since 1960, while the thickness of the ice has fallen by 40 per cent. In 2007, the sea ice around Antarctica had melted back to a record low. At the same time, the movement of glaciers towards the sea has speeded up. A satellite survey between 1996 and 2006 found that the net loss of ice rose by 75 per cent. Temperatures in western Antarctica have increased sharply in recent years, melting ice shelves and changing plant and animal life on the Antarctic Peninsula. Ice melting could cause sea levels to rise by a further 5 m (on top of thermal expansion). Hundreds of millions of people live in coastal areas within this range.

Another consequence is increasing acidity in oceans. As carbon dioxide levels rise in the atmosphere, more of the gas is dissolved in surface waters creating carbonic acid. Since the start of the Industrial Revolution, the acidity of the oceans has increased by 30 per cent. This is having a significant impact on coral reefs and shellfish.

Two relevant, but distinctly different consequences selected. Shows clear understanding of both processes, supplemented with relevant data.

b The main cause of soil degradation is the removal of the natural vegetation cover, leaving the surface exposed to the elements. Deforestation and overgrazing are the two main problems.

Deforestation occurs for a number of reasons, including the clearing of land for agricultural use, for timber, and for other activities such as mining. Such activities tend to happen quickly whereas the loss of vegetation for fuelwood, a massive problem in many developing countries, is generally a more gradual process. Overgrazing is the grazing of natural pastures at stocking intensities above the livestock carrying capacity. Population pressure in many areas and poor agricultural practices have resulted in serious overgrazing. This is a major problem in many parts of the world, particularly in marginal ecosystems. Agricultural mismanagement is also a major problem due to a combination of lack of knowledge and the pursuit of short-term gain against consideration of longer-term damage. Such activities include shifting cultivation without adequate fallow periods, absence of soil conservation measures, cultivation of fragile or marginal lands, unbalanced fertiliser use and the use of poor irrigation techniques.

Topic 4.1

1 a 414094
  b 427098
2 a 4209
  b 4309
3 a north
  b south-south-east
4 a 1 km
  b 2.5 km

5

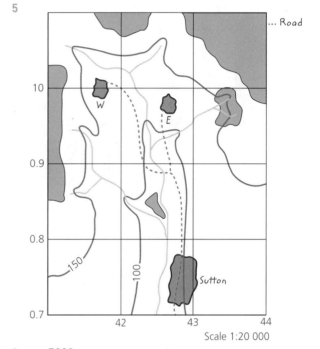

Scale 1:20 000

6 a 5800 m
  b 5895 m
7 4 km
8 4000 m
9 In square 1658 the contours are much closer together than elsewhere, suggesting that it is much steeper. It is generally less steep nearer the icefield and the Karanga Valley.

10

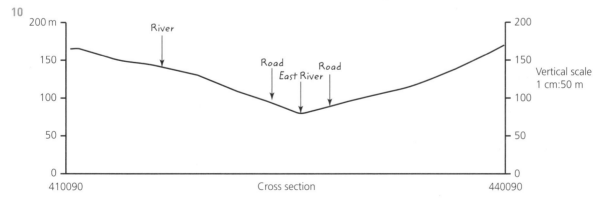

11

2015

☐ More than $1.25/day
■ Less than $1.25/day

12

13

Scatter graph to show
the relationship
between GNI ($)
and life expectancy
(years)

14

Flood plain

River cliff
on outer
bank of
meander –
evidence
of erosion

Meandering
section of river

Straight
section of river

Slip-off slope on inner
bank of meander –
evidence of deposition

Meandering section

Index

Acknowledgements

Every effort has been made to trace all copyright holders, but if any have been inadvertently overlooked, the Publishers will be pleased to make the necessary arrangements at the first opportunity.

p.3 Fig 1.1 *World Population Prospects: The 2004 Revision, 2005* (United Nations/Population Reference Bureau, 2005/2006) © 2005 United Nations. Reprinted with the permission of the United Nations; Fig 1.2 Carl Haub and Toshiko Kaneda, 2012 *World Population Data Sheet* (Washington, DC: Population Reference Bureau, 2012). Reproduced by permission. All rights reserved; **p.4** Fig. 1.3 A Palmer and W Yates, Edexcel (A) *Advanced Geography*, (Philip Allan, 2005); **p.7** Fig 1.5 Population Reference Bureau (PRB), 2011 Kenya Population Data Sheet (Washington, DC: PRB, 2011). Reproduced by permission. All rights reserved; **p.8** Fig. 1.7 from *Advanced Geography: Revision Handbook* by G Nagle and K Spencer (1996) © Garrett Nagle and Kris Spencer 1996. Oxford Publishing Limited. Reproduced with permission of the Licensor through PLSclear; **p.9** Fig. 1.8 *Regional Population Growth, Australia (3218.0)*, Australian Bureau of Statistics © Commonwealth of Australia; **p.11** Fig. 1.9 Source: Earth Policy Institute ® Rutgers University, 2001; **p.13** Fig. 1.10 from *Advanced Geography: Revision Handbook* by G Nagle and K Spencer (1996) © Garrett Nagle and Kris Spencer 1996. Oxford Publishing Limited. Reproduced with permission of the Licensor through PLSclear; Fig. 1.11 *International Migration Report 2015*, UN; **p.15** Fig 1.13 A graph showing the 'Increase in the Mexican-born population in the US from Net Migration from Mexico Falls to Zero–and Perhaps Less', Pew Research Center, April, 2012; **p.16** Fig. 1.14 from *Advanced Geography: Revision Handbook* by G Nagle and K Spencer (1996) © Garrett Nagle and Kris Spencer 1996. Oxford Publishing Limited. Reproduced with permission of the Licensor through PLSclear; **p.17** Fig. 1.15 *CIA World Factbook*; **p.22** Fig. 1.17 from *AS Geography: Concepts and Cases*, P Guinness and G Nagle, Hodder Murray, 2000; **p.35** Fig 2.1 Curriculum Press Limited, Wolverhampton, UK; **p.41** Fig. 2.3 from *AS and A2 Geography for Edexcel B* by G Nagel (2003) © Garrett Nagle, Oxford Publishing Limited. Reproduced with permission of the Licensor through PLSclear; **p.47** Fig. 2.7 from *GCSE Through Diagrams (ORG)* by G Nagle (1998) © Garrett Nagle 1998. Oxford Publishing Limited. Reproduced with permission of the Licensor through PLSclear; **p.48** Fig. 2.9 From *Coral Reefs: Ecosystem in Crisis?* By Sue Warn and Carol Roberts. Field Studies Council, 2001; **p.53** Figs. 2.11 and 2.12 © Garrett Nagle; **p.54** Fig. 2.13 © Garrett Nagle; Fig. 2.15 Source: *Physical Geography in Diagrams 4/e*, Ron Bunnett ISBN 9870582225077, Pearson Education Limited ©1998. Reprinted with permission; **p.55** Figs. 2.18 and 2.19 © Garrett Nagle; **p.66** Fig. 3.3 Source: Paul Guinness, *Geography for the IB Diploma: Patterns and Change*, Cambridge University Press, 2010. Reproduced with permission of the Licensor through PLSclear; **p.68** Fig. 3.4 Source: Paul Guinness, *Geography for the IB Diploma: Patterns and Change*, Cambridge University Press, 2010. Reproduced with permission of the Licensor through PLSclear; **p.72** Table 3.5 Table from the costs and benefits of globalisation in the UK from *Geography for the IB Diploma Global Interactions* by Paul Guinness, Cambridge University Press, March 21, 2011; **p.78** Fig 3.9 Curriculum Press Limited, Wolverhampton, UK; **p.80** Fig. 3.10 from D Waugh, *The New Wider World, 3rd Edition* (Nelson Thornes, 2003). Oxford Publishing Limited. Reproduced with permission of the Licensor through PLSclear; **p. 82** Fig. 3.11 from D Waugh, *The New Wider World, 3rd Edition* (Nelson Thornes, 2003). Oxford Publishing Limited. Reproduced with permission of the Licensor through PLSclear; **p.88** Fig 3.13 Source: *Global Insight: Tourism Satellite Accounting*; **p.92** Fig 3.15 *BP Statistical Review of World Energy, 2017*; **p.94** Fig. 3.17 *BP Statistical Review of World Energy, 2017*; **p.98** Fig. 3.18 *BP Statistical Review of World Energy, June 2013*; **p.100** Fig. 3.19 Source: Paul Guinness, *Geography for the IB Diploma: Patterns and Change*, Cambridge University Press, 2010. Reproduced with permission of the Licensor through PLSclear; **p.106** Fig. 3.21 M Raw, *OCR A2 Geography* (Philip Allan Updates, 2009); **p.107** Fig. 3.22 Source: Paul Guinness *Geography for the IB Diploma: Patterns and Change*, Cambridge University Press, 2010. Reproduced with permission of the Licensor through PLSclear; **p.115** Fig. 4.5 Map image provided by Harms Verlag; **p.116** Fig. 4.6 Extract from 1:25,000 section of French IGN Top 75 Tourism et Randonnée; map Chaine des Puys Massif du Sancy, Institut Geographique National www.ign.fr ISBN 978-2-7585-2720-6 © *IGN – 2019 – Autorisation n 80-1904 – Reproduction interdite.*